THE BODY ELECTRIC

James Geary

THE BODY ELECTRIC

AN ANATOMY OF THE NEW BIONIC SENSES

Rutgers University Press
New Brunswick, New Jersey

First published in the United States 2002
by Rutgers University Press, New Brunswick, New Jersey

First published in Great Britain 2002
by Weidenfeld & Nicolson

The Orion Publishing Group Ltd.
Orion House
5 Upper Saint Martin's Lane
London, WC2H 9EA

Library of Congress Cataloging-in-Publication Data and British Library
Cataloging-in Publication Data are available upon request.

ISBN 0-8135-3194-2

Printed by Butler & Tanner Ltd,
Frome and London

CONTENTS

LIST OF FIGURES

ACKNOWLEDGEMENTS

This book evolved from an article I wrote in a special issue of *Time* called 'The New Age of Discovery' that was published in 1997. I am indebted to my editors at that time, Christopher Redman and James O. Jackson, for giving me the opportunity to edit that special issue and to explore the new bionic senses in my article. I also wish to thank my current editors at *Time*, Ann and Donald Morrison, and *Time*'s managing editor in New York, James Kelly, for permission to update and adapt portions of some of my other *Time* articles on related topics.

Because this is a book about the senses, not a subject that lends itself well to abstraction, I have tried to make the narrative as personal as possible. Wherever practical, I have either experienced the technology I am writing about myself or interviewed someone who has. I've sampled computer-generated smells in London, palpated a virtual liver in Cambridge, Massachusetts, had images projected directly onto my retina in Los Angeles, chatted with intelligent agents on the Web, frolicked with artificial life forms in Cambridge, England and played with about half a dozen robots in Tokyo. I haven't undergone neuro-surgery or any other complex and dangerous operations, however, so when it comes to visual prosthetics and brain implants I have talked to people who have experienced these procedures to find out what it's like to have sophisticated hardware installed in your body. I am grateful to all of them for taking the time to show me their work and talk about their experiences.

I am also grateful to the dozens of scientists, researchers and other individuals whom I've interviewed for this book for their willingness to talk to me (sometimes at great length) about their work and its implications. I owe a particular debt to Rahul Sarpeshkar of MIT, who patiently and repeatedly explained neuromorphic chips to me until I understood them well enough to explain them to others.

My thanks also go to Robert Gesteland, Linda Hoetink and Michael Korkin, who read all or parts of the manuscript at various stages of its preparation. Special thanks go to Andrew Glassner, whose comments on the 'Touch' and 'Hearing' chapters were meticulous and incisive, and to Samantha Hill, who read the entire book and provided many helpful suggestions. The comments and corrections of each of these readers saved me from many mistakes and greatly improved the final text. Any remaining flaws are, of course, strictly my own.

I am also grateful to my editor at Weidenfeld & Nicolson, Peter Tallack, for rejecting my first book proposal (yet another tome on the brain), encouraging me to come up with another idea, and then shepherding that idea into print. The manuscript was greatly improved by his suggestions, particularly in the chapter on 'Sight'.

Finally, I want to thank my sons, Gilles and Tristan, for keeping my senses working overtime all the time. And my wife Linda – *anders had het geen zin*.

London, November 2001

'The senses are the beginning and the end of human knowledge.'
MICHEL DE MONTAIGNE

ONE

THE SILICON SENSORIUM
An Introduction

'It can also be maintained that it is best to provide the machine with the best sense organs that money can buy, and then to teach it to understand and speak English.' ALAN TURING

At the Roman Catholic primary school I attended a lot of the seven-year-olds, myself included, had taken to misspelling the name of the third person in the Trinity. Instead of 'God the Father, God the Son and the Holy Spirit', we would write 'God the Father, God the Son and the Holy Sprit'. This really annoyed the Sisters of Saint Joseph and they drilled the correct spelling into us. 'It's the Holy S-p-i-r-i-t,' they admonished, 'not the Holy Sprit.'

I realise now that the nuns were wrong. If there is anything holy in us, anything that imbues our bodies and minds with a vital spark, 'sprit' is exactly the right word for it.

The term is derived from the Dutch word *spriet*, which means sprout, the first living shoot to emerge from a young plant. From there, sprit's meaning was broadened to denote the spar that juts out diagonally from a ship's mast to extend the surface area of a sail, as in 'bowsprit'. Sprit is, of course, also related to 'sprite' – meaning an elf, pixie, fairy or goblin – a term taken from the French *esprit*, which, in turn, is taken from the Latin *spiritus*, which brings us back to the Sisters of Saint Joseph and their insistence on correct spelling and theological orthodoxy. But for me sprit is both the scientifically more accurate and the metaphorically more compelling term.

Sprit is scientifically more accurate because it's a neat description of a neuron, the nerve cells that conduct the chemical and electrical traffic inside our bodies. Neurons are like tiny sprouts, buds springing from the tendrils of the central nervous system. From these sprouts blossom the five senses – sight, hearing, smell, taste and touch – plus what I call the sixth sense: mind.

Neurons are made up of a central cell body and two main branches: axons, which carry signals away from the neuron to other neurons, and dendrites, which receive incoming signals. Neurons are found throughout the central nervous system – the average human brain contains about 100,000 million of them – and each one is connected to many thousands of others through an intricate lattice of axons and dendrites.

The place where axons and dendrites meet is called a synapse, a tiny gap between neurons through which electrical and neurochemical messages are transmitted from one cell to another. When neurons are stimulated by sensory input, electrical charges are fired across the synaptic gaps from the tips of the axons to the tips of the dendrites. This fusillade is laden with a cocktail of chemicals known as neurotransmitters, which influence the behaviour of other neurons by being absorbed by the dendrites and passed back down the line to the cell body. Once such a message arrives, the receiving neurons fire off signals of their own to other neurons.

The brain alone has at least 10 trillion synapses, and at any given moment trillions upon trillions of impulses are streaking through our skulls, kindling this elaborate neural network into a flurry of signalling activity. The barrage of neural firing caused by these electrical and chemical signals forms the basis of all our perceptions, thoughts, emotions and memories. This is how neurons take the raw data of the world out there and translate it into the world within.

Apart from its scientific accuracy, sprit is the metaphorically more compelling term because neurons catch the whirlwind of sensation that is the world – its profusion of colours, sounds, scents, flavours, textures and impressions – and use it to propel the body and mind. The senses are so essential to our navigation of reality – there can be no 'our' and no 'reality' without them – that language is steeped in their metaphors. We understand someone when we 'see' what she means or 'hear' what she's saying; a dubious scheme doesn't 'smell' right; an unpleasant experience leaves a 'bad taste' in your mouth; a sensitive person is 'in touch' with his feelings. We can only comprehend and engage the world when the transaction is phrased in the body's own vocabulary.

That vocabulary is now being expanded, as a bowsprit extends the surface of a sail, by connecting sprits and chips. Human beings are embedding computer chips inside their bodies to enhance or augment their senses, while computers are increasingly being given the ability to see, hear, smell, taste and touch. And once a computer has its own silicon sensorium, it's conceivable that it could at some point learn to think.

This remarkable convergence of body and machine is being brought about by wedding advanced computing technology to the human nervous system, a marriage that holds the promise of devices that can restore useful sight to the blind and help victims of paralysis to regain partial use of their limbs. Brian Holgersen – a thirty-year-old Danish tetraplegic whom we'll meet in the chapter on 'Touch' – is a case in point. Holgersen was paralysed from the neck down after a motorcycle accident, but thanks to electronics implanted in his body he can now hold a cup, lift a fork and grasp a pen, actions he was previously unable to perform.

This mixture of flesh and chips could also one day give individuals so-called bionic senses, such as the ability to see infrared radiation or to feel objects at a distance. Some people, such as Kevin Warwick, a professor of cybernetics at the University of Reading in the United Kingdom, even suggest that computers will eventually endow the human body with extrasensory perceptions.

In the summer of 1998 Warwick had a small silicon chip implanted in his forearm. The chip transmitted a radio signal through which he was able to operate and interact with a number of electronic devices in his immediate environment. When he arrived at the office, for example, his computer recognised the chip's radio code, unlocked the door, switched on the lights, called up his personal website and greeted him with a friendly 'Hello, Professor Warwick'. The experiment was intended to show how, in Warwick's words, 'Humans and machines can work together, combining the best features of both.'

If a person can communicate with a computer through a silicon chip implanted in the body, Warwick wonders, why not with another person? His next attempt to meld man and machine might provide an answer to that question.

He hopes to link two nervous systems – his own and that of his wife, Irena – via the Internet. This time corresponding computer chips, about the size of postage stamps, will be surgically connected to nerve endings in the couple's arms. In theory, Warwick's chip should pick up the electrical nerve impulses travelling down his arm – when he becomes angry, for instance, or when he's feeling frisky – and transmit them over the Internet to the corresponding chip in his wife's arm.

The result: who knows? Since such an experiment has never been done before, it's impossible to guess what kind of message – if any – will be exchanged. 'If I'm feeling angry,' Warwick says, 'will that anger get blasted down onto Irena, or just some vague tingling sensation in her arm?' Warwick is clear about one thing, though: if this digital mind-reading works, the ultimate application will be 'telepathy through the Internet'.

Telepathy through the Internet is a long way off, if it's even possible at all. One major difficulty with Warwick's idea is that thoughts and emotions are not communicated by electrical impulses alone; neurochemicals and hormones play an equally important role, yet Warwick's scheme makes no provision for transmitting these.

Nevertheless, people like Warwick and Holgersen are living examples of the way computers are getting under our skins. They are cyborgs. The term, a conflation of the words 'cybernetic' and 'organism', was coined in 1960 by Manfred Clynes and Nathan Kline in an article called 'Cyborgs and Space' published in *Astronautics* magazine. Cybernetics, the study of society through an analysis of the messages exchanged 'between man and machines, between machines and man, and between machines and machines', was launched in the 1950s by computer pioneer Norbert Wiener.

When Clynes first came up with the term 'cyborg', Kline quipped that it sounded like a town in Denmark. They intended the word to denote a human being outfitted with technological enhancements that alter 'man's bodily functions to meet the requirements of extra-terrestrial environments'. (The word 'bionic' was invented two years earlier, in 1958, also to describe the technological enhancement of the body.) The authors argued that to make space travel possible, a closer integration of man and machine was necessary, a merger so intimate that machines would gradually come to be regarded as essential parts of the human body. Clynes originally meant the concept to include anyone who integrated technology with his or her physiology, including a person riding a bicycle, wearing glasses or listening to music through headphones. It was only later, as a result of films like *Terminator*, that the term took on its present somewhat sinister connotations. Today, cyborgs are altering their bodily functions not for space travel, but to meet the requirements of life here on Earth.

Like all new ideas, the basic concepts behind cyborgs have a long and distinguished history. As far back as 1665, English physicist and mathematician Robert Hooke in his work *Micrographia* wrote, 'The next care to be taken, in respect of the Senses, is a supplying of their infirmities with Instruments, and as it were, the adding of artificial Organs to the natural ... and as Glasses have highly promoted our seeing, so 'tis not improbable, but that there may be found many mechanical inventions to improve our other senses of hearing, smelling, tasting, and touching.'

This book is about the people who are making and using the 'many mechanical inventions' to improve, enhance or repair the senses. The book explores the convergence between biology and technology in each of the five senses plus the sixth sense, mind. In the chapter on 'Sight', for example, we'll meet Marie, a sixty-three-year-old Belgian woman, who has been totally blind since the age of fifty-seven. Marie is one of only about a dozen people in the world who has a visual prosthetic; in Marie's case, it's an electrode array implanted around her right optic nerve that enables her to see light, shapes and colours again. In the chapter on 'Touch' Miguel Nicolelis, associate professor of neurobiology at Duke University Medical Center in North Carolina, describes his work on neuron–silicon interfaces in the brains of monkeys, and how one day this technology could give people extra virtual eyes, ears and limbs controlled through the Internet. In the chapter on 'Mind', Japanese researcher Keiichi Torimitsu explains how he's trying to replicate the brain's visual processing by growing rat neurons on a silicon chip.

While scientists such as Nicolelis and Torimitsu are working to meld technology with biology, other researchers are trying to endow computers with human-like sensory abilities. The chapter on 'Sight' details how computers can not only see, but can also detect whether you're telling the truth through an analysis of your facial expressions. The chapters on 'Smell' and 'Taste' describe how companies have developed devices to transmit odours and flavours over the Internet and how these sensual delights could be used to entice online consumers. The chapter on 'Touch' explores the realm of haptic interfaces, systems that enable users to reach in and feel the information inside computers.

I think it only fair to warn readers up front: this book is not about how human beings are turning into computers or how computers are turning into human beings. People are not computers and computers are not people. And while it's a fact that humans and machines are becoming entwined in an ever more intimate embrace, that's not the same thing as the one becoming the other. Millions of years of evolution have made people the way they are, and I think it very likely they will remain that way – regardless of how many chips are implanted in their bodies. And while it's entirely possible that computers might someday evolve into entities recognised as viable life forms in their own right, that doesn't make them human. If such an evolutionary process does take place, it will take place as it has for countless other species: according to the unique demands of the computers' physiology and environment. It's possible that computers will evolve into living machines, but not into human beings.

While this book is all about the senses, I would hope that it is not sensationalistic. I have endeavoured to stay close to the science throughout, and all the technologies described herein are actually in use. Some are still in the earliest experimental stages; others have already been commercialised. None, however, is yet widely used outside some very specific laboratory or industrial applications. When I do speculate about the future – about the possibility of having virtual senses through the Internet, for example – my musings are grounded in proven research that is going on right now.

There are, of course, other routes than through silicon to augment, enhance or extend the senses; biotechnology, nanotechnology, genetics and tissue engineering are just a few. There are also other ways in which technology is being incorporated into the human body, such as the artificial heart made by AbioCor. And there could well be more senses than the six described in this book. Some researchers suggest, for example, that the vomeronasal organ, which is located in the nose and is believed to detect odourless pheromones, is an additional sense that can influence mood and behaviour. These subjects are beyond the scope of this book, because either they do not deal primarily with the senses (as with the artificial heart) or they do not deal primarily with computing technology (as with genetics and the vomeronasal organ). I have limited my focus strictly to the confluence of sprits and chips.

'With the arrival of electric technology, man extended, or set outside himself, a live model of the central nervous system itself,' Marshall McLuhan wrote in *Understanding Media*. The television is an extension of the eye, the telephone an extension of the ear. Through the creation of silicon senses, these extensions are being turned back in on themselves – the electric technologies that have extended our senses around the world and into outer space are now being integrated into our bodies to extend the senses even further. At the very least, this will lead to a much more direct and intimate relationship between our bodies and our machines. Eventually, our senses may not merely be extended: we could end up having more of them.

The subtitle to McLuhan's book, in which he anticipates many of these issues, is often forgotten: *The Extensions of Man*. 'Any extension,' he notes, 'whether of skin, hand, or foot, affects the whole psychic and social complex.' Because the merger of man and machine affects more than just biology and technology, I have tried to address some of the psychological, sociological and philosophical implications of this work. Are you any less 'you' after a bionic implant? If all our senses are enhanced, how will we tell the difference between virtual reality and the actual world? Will it matter? How can privacy be ensured when computers are watching and listening to everything we do and say? Will transmitting smells and tastes over the Internet enrich the user's experience or merely provide another way for corporations to sell us stuff? These are just some of the ways that 'the framework itself changes with new technology, and not just the picture within the frame', as McLuhan put it.

This book chronicles how that framework is changing as the human body becomes electric. From these first grafts between sprits and chips, who knows what senses might spring? The silicon sensorium is already changing the way we see, hear, smell, taste, touch and think about the world, opening up the doors of perception just another crack. One day the new bionic senses might blow those doors completely off their hinges.

TWO

SIGHT

'Who are you going to believe – me or your own eyes?' GROUCHO MARX

'It was wonderful, extraordinary,' says Marie, a sixty-three-year-old Belgian woman, of her experience in a tiny room on the second floor of the University of Louvain in Brussels.

'It was fantastic,' says David Robinson, a twenty-nine-year-old student of computer networking, of his afternoon on an operating table at the Massachusetts Eye and Ear Infirmary in Boston.

What has moved these two people to such enthusiasm? In Marie's case, it was a vision of red, blue and yellow dots arranged in neat little rows like the tops of Lego building blocks. For David, it was the sight of bright circles of light and long skinny lines that resembled different shapes and sizes of 'peas and pencils'.

To most people, glimpses of Lego bricks, peas and pencils are hardly worth getting excited about; but Marie and David are blind. They suffer from retinitis pigmentosa, a degenerative eye disease in which the photoreceptor cells in the retina gradually die off while the rest of the eye remains healthy. Marie has been functionally blind – a condition in which only the ability to perceive light and shadow and some vague shapes is retained – since the age of forty and completely blind since the age of fifty-seven. David has been functionally blind since the age of thirteen.

Marie (not this woman's real name; she wishes to remain anonymous) can see again, thanks to an electrode implanted around her right optic nerve. The electrode is connected to a stimulator permanently installed in a small depression carved from the inside of her skull. A video camera, worn on a cap, transmits images in the

9

form of radio signals to the stimulator, which converts these signals into electrical impulses and sends them along Marie's optic nerve. The optic nerve ferries the signals to Marie's visual cortex, where they are reassembled into an image: in this case, a collection of red, blue and yellow Lego bricks. 'The device is an integral part of my body,' Marie says. 'I don't feel it. I completely ignore it.'

David's sight was partially restored only once, during a six-hour operation at the Massachusetts Eye and Ear Infirmary. In David's case, the electrical current was supplied by a simple switchbox and the experiment showed that his retina, the thin sheet of light-sensitive cells stretched across the back of the eye, still retained some residual visual capacity that could be used to restore useful sight. David's implant, which was removed immediately after the experiment, is still in the earliest stages of development, but the final device will consist of a silicon microchip attached to the surface of the retina. The microchip will eventually function like Marie's stimulator, converting visual signals from a video camera into electrical impulses and transmitting these to the optic nerve. In both techniques, the signals bypass damaged photoreceptors to convey visual input to the brain.

Normally, David is unable to see clear shapes. To him, the edge of a table – something he can only see if it's no more than a metre away from him – appears as an indistinct blob made up of a mass of broken lines. Although not everything he saw during the procedure was equally distinct – some percepts were even more blurred than what he sees unaided – he was able to perceive basic shapes again. 'I haven't seen a straight line in thirteen years,' he says, 'so I didn't know what they looked like any more. It was fun being able to see shapes, circles and angles again.'

Marie and David are two of only about a dozen people in the world who have had visual implants that could potentially restore their sight. Experiments are currently under way in Europe, Japan and the United States to install microelectronic arrays in the retina, around the optic nerve and even in the visual centres of the brain itself to restore partial vision to people suffering from blindness due to diseases like retinitis pigmentosa. The implants cannot yet restore full sight to the blind, but researchers hope that they will eventually help patients distinguish between light and dark and discern the outlines of objects and basic shapes. Some suggest that, as the technology improves, these devices might one day even provide users with enhanced visual acuity, such as the ability to see things normally invisible to the human eye. Before that happens, though, researchers will have to prove that these devices produce useful ordinary vision and are compatible with the body's organic tissues – and that the long-term benefits outweigh the immediate surgical risks.

After more than twenty years of functional blindness and almost seven years of total blindness, Marie isn't looking for any kind of exotic augmented vision. Ordinary vision would suit her just fine, and she's delighted with her newly restored ability to see basic shapes, shadows and colours. She started to lose her sight when retinitis pigmentosa caused the rod and cone cells in her retina to degenerate. The human retina is made up of about 120 million rod cells, receptors that operate under relatively dark conditions, are highly sensitive to movement but do not respond well to colour. The retina also has seven million or so cone cells, which function under brighter conditions, are sensitive to colour and provide high spatial detail. Cone cells come in three different types: one that responds to red light, another that responds to green light, and a third that responds to blue–violet light. The rod cells provide the brain with information primarily about where things are, while the cone cells tell the brain what things are.

In people with retinitis pigmentosa, as well as a similar condition known as age-related macular degeneration (AMD), the rod and cone cells die off, rendering the retina insensitive to light. Marie's case followed the typical pattern. First her rod cells went, depriving her of most of her peripheral sight and leaving only a slim tunnel of vision through which she could still manage to recognise objects and read with difficulty. But then as her cone cells failed, this last narrow window on the world snapped shut and she was left completely blind, despite the fact that her retina retained a healthy link to the visual centres of her brain through a functioning optic nerve. Marie's implant splices into the live line of the optic nerve to enable her to see again.

Visual images are perceived when rays of light, made up of particles called photons, enter the eye through the pupil, pass through the cornea and strike the retina. The rod and cone cells in the retina convert the incoming light into electrochemical signals and send them to ganglion cells, the axons of which make up the optic nerve. The optic nerve then conducts these impulses to the brain. There are different brain centres for different kinds of visual information: information about where an object is, for example, is processed in one part of the brain, while information about what it is is processed in another part. Just as a switchboard automatically routes thousands of telephone calls to the correct extensions, the optic nerve directs millions of visual signals to the appropriate brain areas.

Marie's artificial visual system is called the Microsystem-based Visual Prosthesis (MIVIP) and was designed by Claude Veraart and collaborators at the University of Louvain in Brussels as part of a European Union project. The MIVIP consists of a spiral cuff electrode implanted around Marie's right optic nerve. The electrode, which wraps around the optic nerve like the little plastic sheath on the end of a shoelace, is connected to a thin cable encased in silicon rubber that snakes its way from the optic nerve that exits the back of Marie's eye around the outside of her brain to the stimulator implanted in a small cavity in her cranium. The stimulator, which bypasses the damaged rod and cone cells to send electrical signals directly to the optic nerve, is operated by means of radio signals transmitted from an external video camera.

These devices are permanently implanted in Marie's body, but to use them she must travel to a small room at the University of Louvain and don what looks like a badly damaged 1960s bathing cap. The cap is made of plastic and has a standard video camera affixed to its front. The video camera is linked to an external transmitter, also attached to the cap, that is aligned directly above Marie's implanted stimulator. Marie sits in front of a large white screen on which an alphabet of about fifty different line configurations is projected.

The configurations resemble large letters – an X or an H, for example – basic forms such as blocks and circles or shapes that look like tables and chairs. The projected images contain live pixels that the video camera registers as a flash when it passes over them. As the camera crosses a live pixel, it sends a signal to the transmitter, which passes it on to the stimulator, which sends an electrical charge to Marie's optic nerve. The result: Marie sees a series of flashes, called phosphenes, that join up to form recognisable shapes.

Since the camera's visual range is narrow, Marie has to scan an image by slowly moving her head from left to right and up and down

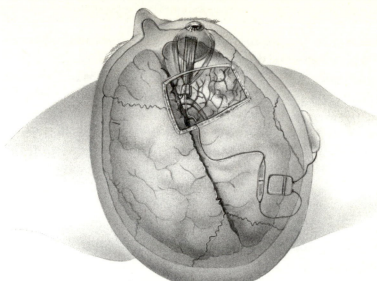

Figure 1. The Microsystem-based Visual Prosthesis (MIVIP), designed by Claude Veraart and collaborators at the University of Louvain in Brussels, consists of a spiral cuff electrode implanted around the optic nerve, which emerges from the back of the eye. The electrode is connected by a thin cable to a stimulator implanted in a small depression in the skull. The stimulator receives radio signals from an external video camera, which are translated into electrical impulses that stimulate the optic nerve.

until she's covered the entire screen. As the camera criss-crosses the visual field, a rapid series of electrical stimulations is sent to her optic nerve. The number of electrical stimulations depends on the number of live pixels on the screen: the more there are, the easier and quicker it is to compile an image. Marie reconstructs the image from what appear to be a series of strobe flashes. As I watched a simulation of how the device works, a string of rapid flashes formed two intersecting diagonal lines. After a few seconds, it was clear that the image was an X. The experience was a bit like watching a miniature stadium billboard, on which images are also compiled from groups of individual flashing lights.

Seeing is a slow process using the MIVIP. It takes Marie on average about a minute of back-and-forth scanning to distinguish a shape. As images flash before her eyes, she is asked to draw what she sees and describe the part of the visual field in which she sees it. In the experiments carried out to date, Veraart has noted a correspondence between where Marie sees a phosphene and that image's actual location in space, which means that the flashes transmitted to Marie's optic nerve correlate with the outside world.

Veraart has also discovered that the phosphenes Marie sees are different depending on the electrical signal's duration, strength and frequency. In other words, different electrical pulses lead to different perceptions. One type of pulse might always produce a yellow phosphene, for example, while another might always produce red. If this turns out to be the case, Veraart and his team intend to compile a lexicon of correspondences so that specific visual stimuli can be easily reproduced.

At this stage, the MIVIP is still a cumbersome and limited device. It cannot give a person back her sight. At best, it can help restore mobility by helping people to avoid obstacles, recognise landmarks in unfamiliar environments and detect very simple shapes. Imagine public spaces seeded with a kind of invisible Braille, live pixels embedded in doors, stairways and street corners that blind MIVIP users could employ to see important information about their immediate surroundings. Veraart and his colleagues are working to refine the technology so that the blind can actually see obstacles such as chairs and tables in this way.

Before that happens, though, the device has to become smaller and more portable. Moreover, the implant only works for patients suffering from retinitis pigmentosa and AMD, in which the optic nerve and visual centres of the brain are left intact. But Veraart argues that visual prosthetics could be especially important for this group, since retinitis pigmentosa and AMD together affect about 30 million people worldwide and are the most common causes of untreatable blindness in developed countries. And because the diseases cause late-onset blindness – Marie was forty when she started to lose her sight – this group often has great difficulty coping with the condition, having been accustomed to independence and mobility all their lives. A little help from the MIVIP would be welcome.

'This is not true vision,' stresses Veraart. 'And it's definitely not a cure for blindness. It's something to help people better cope with their impairment. It's like a wheelchair: it doesn't help people walk again, but it does help them get around. And as a technological solution, that's not bad at all.'

Marie agrees. She has undergone extensive, and dangerous, brain surgery to use the MIVIP and she still shows up every week at Veraart's lab for more tests and experiments. 'Even if I recover only light and shadows,' she says, 'it would still be worth it.'

THE NAKED EYE

The optic nerve is not the only spot at which a visual prosthetic might work. Theoretically, an implant could be installed anywhere along the visual pathway. On the retina itself, there are two possible intervention sites: epiretinal implants (electrode arrays placed on the surface of the retina, which communicate directly with the ganglion cells that link rod and cone cells to the optic nerve) and subretinal implants (which replace the rod and cone cells themselves).

Joseph Rizzo, an associate professor of ophthalmology at Harvard Medical School is co-director (with John Wyatt of the Massachusetts Institute of Technology) of the Boston-based team that performed the experiment on David Robinson. They are developing an interface for the surface of the retina, where vision begins. 'If an electronic implant is placed directly on the retina, it can feed optical signals into the nervous system right at the point where they normally originate, before complex and poorly understood signal processing takes place,' says Rizzo.

Rizzo and Wyatt's epiretinal implant will comprise a laser mounted on a pair of glasses and two silicon microchips connected to a small electronic camera. It will work like a television transmitter for the eye. The camera captures a visual scene and transmits the image to a signal-processing microchip on the glasses. This chip converts the visual information into an electronic code, which is projected by a laser beam through the subject's eye to a stimulator chip resting on the surface of the retina. The stimulator chip decodes the picture information carried by the laser and transmits electrical pulses to the nearby ganglion cells, which forward the signals to the optic nerve and, eventually, on to the brain. The images captured by the camera correspond to the electronic signals transmitted to the optic nerve by the stimulator chip. If the camera captures an image of the letter H, for example, a pattern corresponding to the letter H is activated on the chip.

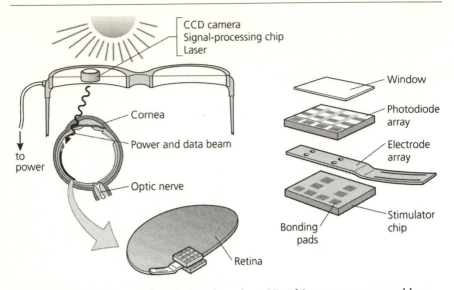

Figure 2. **a**, The Harvard/MIT retinal implant. Visual images are captured by a tiny camera mounted on a pair of glasses. The camera transmits images to a signal-processing microchip on the glasses, which converts the visual information into an electronic code that is projected by a laser beam through the subject's eye to a stimulator chip on the retina.

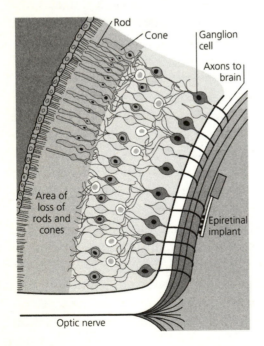

Figure 2. **b**, The stimulator chip decodes the picture information carried by the laser and transmits electrical pulses to the nearby ganglion cells, which forward the signals to the optic nerve and, eventually, on to the visual centres of the brain. The images captured by the camera correspond to the electronic signals transmitted to the optic nerve by the stimulator chip.

In David Robinson's case, those patterns looked like peas and pencils. David was one of the first volunteers in a series of experiments to determine the quality of perception that can be attained through an epiretinal array. During the operation, David received only local anaesthetic because he needed to be conscious to describe – and draw – what he saw.

After the device was implanted, David's right eye was exposed to 255 stimulations. The images he saw were sometimes as dull as soft candlelight and sometimes as bright as a car's headlights – and he drew a lot of pictures of peas and pencils. 'The electrical stimulations felt a bit like prickly needles in my eye,' he says. 'But the pain was manageable.' Despite the risks – the surgery is still experimental, so he could have lost his eye or emerged with even poorer vision – David would recommend the procedure to blind friends. But Rizzo is cautious. 'The blind can be made to see very crude shapes,' he says. 'The technology has great potential but it hasn't been proven yet.'

Alan Chow hopes to prove the technology's potential – and soon. So far Chow, a paediatric ophthalmologist and co-founder of Optobionics Corporation in Wheaton, Illinois, has implanted visual prosthetics under the retinas of three blind patients who lost almost all their vision as a result of retinitis pigmentosa. The surgeries were part of a study approved by the US Food and Drug Administration to determine whether subretinal implants can be tolerated in the human eye.

Chow's implant, called an Artificial Silicon Retina™ (ASR) and developed with his brother Vincent, is a microchip containing some 3500 solar cells, each one in contact with the retina. The solar cells convert light into electrical impulses, which are then used to stimulate the remaining healthy retinal cells.

The surgical procedure to install the implant begins with three tiny incisions in the sclera, the white part of the eye. Then a pinpoint opening is made in the retina through which fluid is injected, creating a small pocket in the subretinal space just wide enough to accommodate the ASR chip. The surgeons then carefully slide the ASR into the pocket, reseal the retina over the chip, introduce air into the middle of the eye to gently push the retina back down over the device, and close the incisions.

Unlike the other prosthetics described thus far, the ASR requires

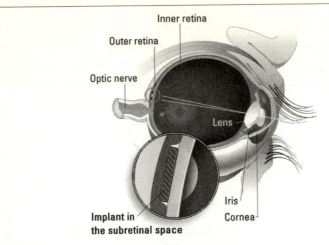

Figure 3. The Artificial Silicon Retina (ASR).

no external technology at all – no video cameras, no stimulators, no batteries – but is powered solely by natural light. When implanted, the ASR is intended to produce electrical signals similar to those produced by the photoreceptor cells in response to actual light, not encoded video images, from the environment. 'We want to produce not only the sensation of light and darkness, but also the sensation of different intensities of light,' says Chow. 'The best chance of doing this is in the subretinal space.'

There are several technical obstacles to be overcome before implants like the ASR can be approved for use in humans. First, the implants must continue to function and remain in stable positions inside the patient's eye. And the patients must be able to tolerate the device without discomfort, infection or rejection. If the ASR passes these tests, Chow speculates that it could be possible in the future to integrate other devices into retinal implants. X-ray vision will likely remain in the realm of science fiction, but other types of enhanced sight, such as infrared vision, might be possible. 'Silicon responds to near infrared wavelengths, so we might be able to see things beyond what we can see now,' says Chow. 'We won't be able to see through walls or clothes, but if a car exhaust is hot, for example, a person might be able to see an infrared glow around it.'

Though they haven't yet managed to produce infrared vision, researchers at the Dobelle Institute in New York claim they have enabled a sixty-two-year-old blind man to navigate the New York subway system, thanks to an electrode array implanted in his brain. The Dobelle eye is similar to Rizzo and Wyatt's device. A television camera and an ultrasonic distance sensor, both of which are mounted on a pair of eyeglasses, are connected to a portable computer worn on the subject's belt. The computer processes the video and distance signals, converts them into electrical pulses and transmits them to the electrodes implanted in the man's visual cortex. Like David Robinson, the subject sees flashes of light and luminous shapes that correspond to the image transmitted from the video camera. Dobelle researchers say the man is learning to use his artificial vision to watch television and operate a computer through a special interface. The Dobelle work also suggests that biocompatibility may not be a problem: the man in question had his implant back in 1978 and has been walking around with the device in his head ever since.

Biological systems excel at sensory perception because brains have evolved the ability to process enormous amounts of information quickly, efficiently and inside very constrained spaces – even when that information comes from electronic devices. Brains, however, are not computers and they don't make sense of the world purely as strings of binary signals (also known as bits), as digital devices do. When dealing with input from the senses the brain performs like an analogue device, meaning that it processes information that changes over time. So the brain is both analogue and digital at the same time.

The electrical discharge that takes place when a neuron fires is called an action potential, or spike. Spikes are digital, all-or-nothing pulses. The time interval between spikes, however, is analogue because any continuous time interval is possible, depending on what else is going on in the brain. Because the time between spikes is analogue while the spikes themselves are digital, the brain's signal representations are not purely one or the other but a hybrid of the two.

The spike firing pattern of any given neuron in the brain is primarily determined by the spike firing patterns of the thousands of other neurons to which it is connected; but the spike firing patterns of neurons that receive analogue signals from the senses are primarily determined by sensory input, and sensory input is analogue because it is in a constant state of flux.

When you see a face, for example, your retina converts analogue light waves into the spike signals of the nervous system. The brain makes a digital decision: the object before you either is or isn't a face. But that decision is based on analogue processing because the face in question is changing even as you look at it: patterns of light and shadow are shifting; facial expressions appear and fade and gradually morph into different ones; even your perspective changes as you see the face first from the front and then from the side. Despite this jumble of data, the brain still detects a face. And not just any face, but the face of a particular friend, family member or total stranger.

People are good at recognising faces because they have to be – survival can depend upon it. Newborns are able to distinguish faces from other objects within about forty-five minutes of birth, and within two days express a marked preference for their mother's face over those of strangers. Computers find facial recognition a lot more difficult, though, partly because the human face is such a complex visual stimulus. Machines often have trouble detecting whether a face is even present in a given stretch of video tape, much less if that face is gasping in horror or chortling with delight. For humans, this same task is kid's stuff.

Face perception, whether by people or computers, depends on pattern recognition – spotting the shapes and configurations that make up a human countenance. Most recognition systems scan stretches of video tape to find these recurring patterns; but to analyse the patterns they rely on neural networks, computer programs that operate in a manner analogous to the way neurons work in the brain. Neural networks are a crucial part of many of the bionic technologies discussed in this book.

Neural networks are important because they emulate the brain's pattern-recognition skills. A neural network is built up out of layers, like lasagne. Each layer consists of a group of processing elements known as 'artificial neurons'. One layer of artificial neurons is for input, mostly for receiving information from the environment; another layer is for output, mostly for communicating the network's response to a particular stimulus. The layers in between are where most of the information processing happens.

Learning takes place in the human brain when new information is stored in the form of strengthened connections among a select group of neurons. When a baby learns to recognise its mother's face, for example, a specific network of neurons in the visual centres of its brain will fire in unison. The more often the baby is exposed to its mother's face, the more often this network fires. The connections among the neurons strengthen each time and so the learning is reinforced.

Artificial neurons work the same way – they learn from experience. Neural networks are not programmed with a knowledge of faces. Instead, they are exposed to a variety of visual stimuli and learn to sort the faces from the non-faces themselves. This is done through a process called 'back propagation'.

In back propagation, the network is exposed to visual inputs – say, things like faces, household objects, straight lines and rounded surfaces. As it performs its analysis it is graded by a teacher on how well it is doing. The teacher can be an actual person or another computer program that already knows the right answers. Then information about the machine's errors is also filtered back into the system to help improve performance. As the network becomes more proficient at distinguishing faces from tables and chairs, it stores this new knowledge in a pattern of strengthened connections among its artificial neurons – which brings us back to neuromorphic chips and their ability to replicate the brain's analogue and digital behaviour. When part of a neural network, such chips have promising applications for the processing of sensory information, especially visual or auditory information.

'The average human brain consumes just 12 Watts of power – one-tenth of what it takes to burn an ordinary light bulb – and computes in real time in an area not much bigger than two slices of pizza,' says Rahul Sarpeshkar of the Massachusetts Institute of Technology, one of the scientists who is developing neuromorphic chips. Neuromorphic chips contain circuits that are inspired by neurobiological systems, but morph these systems so that they are more suited to silicon. According to Sarpeshkar, the best way to match what neurobiological structures such as the eye, ear and brain do is to try to replicate their key computational ideas in electronics. So neuromorphic chips are being investigated for use in sensory data processing applications such as speech recognition, in robotics and in the creation of artificial eyes and ears.

Writing in 1849, English physician Alfred Smee anticipated the development of artificial eyes, describing something that comes very close to what we now know as television:

> From my experiments I believe that it is sufficiently demonstrated that the light falling upon the [optic] nerve determines a voltaic current which passes through the nerves to the brain. From this fact we might make an artificial eye, if we did but take the labour to aggregate a number of tubes communicating with voltaic circuits ... Having one nervous element, it is but a repetition to make an eye; and ... there is no reason why a view of St Paul's in London should not be carried to Edinburgh through tubes like the nerves which carry the impression to the brain.

Artificial eyes like those described by Smee already exist, of course. You might even have one lying around the house. A digital camera is a kind of artificial eye that uses a charge-coupled device (CCD), a tiny array of millions of light-sensitive transistors, to transform photons into electrical signals. The brighter the light that hits an individual transistor, the greater the electrical charge created in a 'bucket' under the transistor. The bucket is relatively empty of electrical charge until light induces charged particles to start filling up. To create an image, an analogue-to-digital converter translates these analogue electrical signals into a series of digital values. The camera then reproduces the image from these values, which, taken as a whole, are a close approximation of the original pattern of light that struck the array.

While a CCD takes a pretty good picture, it's not much use when it comes to vision. Ironically, that's not because it doesn't 'see' enough but because it sees too much. A single second of video tape contains about 22 megabytes of data, the very rough equivalent of about thirty copies of this book. But most of that data does not convey information that an organism is interested in. Indeed, the brain would quickly be swamped if it had to process that much visual information every second. Instead, the brain starts editing right away. As visual information proceeds from the eyes through the optic nerve to the brain, a little bit of information is selectively destroyed at each stage along the way. Each level of processing acts as a kind of filter, enabling a very complex image to be pared down to its most important elements.

Like the visual system of the brain, neuromorphic chips preprocess this barrage of visual input to pick out the important bits and discard the rest. CCDs can't perform this trick. 'In vision, whether it's biological or artificial, you want to recognise patterns,' Sarpeshkar says. 'To do that, you don't need all the information in an image. You just need the important information. So the best strategy is to do analogue processing right there where the information is, and transform this information into a less overwhelming and more useful form before shipping it to the brain.'

The photoreceptors on a neuromorphic chip do this by sending an analysis of an image to the processing unit, not the entire image itself. A CCD, on the other hand, has to do this analysis somewhere else: the raw data is sent to an analogue-to-digital converter, and reams of digital data then have to be processed and analysed before any meaningful information can be extracted from the image. Extensive digital processing takes a long time, eats up a lot of power, and makes CCDs unattractive for use in computer vision applications that need to be portable and have to operate in real time. That pretty much rules them out for use in robotics and prosthetics.

Neuromorphic chip emulations of biological vision systems rapidly and efficiently extract all kinds of useful information from the world. One crucial piece of information is optical flow, a map of the local velocities of objects in an image. When you're gazing out of the window of a moving train, for example, the trees along the tracks flash past in an instant but more distant hills roll slowly by. These varying velocities generate an optic flow map that correlates with the distances of objects from your eye.

Neuromorphic chips detect optical flow in real time, and so could be used in a guidance system to help a robot make its way through its environment. A robot with two vision chips for eyes could use optical-flow information to detect when it was in danger of bumping into an object. If optical flow in its left eye increased relative to its right eye, for example, it would know that it was approaching an object on the left. By keeping the optical flow in both eyes balanced, the robot could successfully cross a room or even run an obstacle course. A robot equipped with CCDs instead of vision chips could be made to perform the same calculations, but the CCDs could take so long to get a result that in the meantime the robot could have bumped into a wall or plunged off the edge of a cliff. Faster CCD implementations will require powerful, power-hungry computers that are awkward for mobile, autonomous robots.

Different kinds of chips are good for different kinds of vision. Sarpeshkar works with chips based mostly on the visual system of the fly, which has evolved to specialise in detecting motion. Flies have compound eyes, so called because they are made up of repeated segments known as ommatidia. Each ommatidium captures visual information from just a single area of the fly's field of view. The information taken in from all the ommatidia together constitutes the image, just as a photo in a magazine is made up of a mass of individual dots. Flies are good at detecting motion because, as an object moves across the visual field, the ommatidia flicker on and off, creating the visual equivalent of a ship's wake in the fly's eye. Chips based on the fly eye would be ideal for the mobile robots described above.

Other chips are based on the human eye, which is more adept at detecting patterns and more suited to applications like face recognition. Though neuromorphic chips are still in the early stages of development, potential applications include things like cars that drive themselves by keeping their own eyes on the road or doors that only open when they recognise a familiar face. One of the most successful neuromorphic devices built to date is a model airplane that flies itself, albeit on a tether, by means of its own on-board vision system.

Once a working electrode–neuron interface is perfected, neuromorphic vision chips and electrodes could combine to substitute for damaged eyes. More fantastic applications could also be imagined, such as the supersight enjoyed by Steve Austin, the bionic secret agent in the 1970s television series *The Six Million Dollar Man*. The sensors in an artificial eye might, for example, be tuned to normally invisible parts of the electromagnetic spectrum, enabling people with neuromorphic eye implants to see in the dark. In fact, the beginnings of bionic vision already exist.

MORE THAN MEETS THE EYE

Picture this. You are walking down a street in a foreign city, wondering how to find a particular restaurant at which you've arranged to meet an old friend. After entering a few commands on a small computer attached to your belt, you see a city map appear before you in the air, with the quickest route to the restaurant outlined in yellow. You then select another option, which points you in the right direction. Having eventually found the street, you can just make out a sign in the distance. A light touch to your glasses magnifies the image, confirming that this is indeed the place. You enter the restaurant and recognise your friend already seated at a table. Now picture one more thing: you are blind.

This is the future of sight envisioned by Tom Furness, director of the University of Washington's Human Interface Technology (HIT) Laboratory in Seattle. For the past thirty years, Furness has been studying how the sense of sight can be enhanced – and, in some cases, even restored – through the use of lasers and advanced computer technology. The result is the virtual retinal display (VRD), a device that paints images directly onto the human retina. 'It's like taking a movie projector and shining it directly onto the retina,' Furness explains. 'Instead of creating a screen to look at, we create the light rays that appear to emanate from that screen. When the light reaches the retina, it appear as if the image is really there – but it's not. It's a virtual image.' When I tried the VRD at a computer conference in Los Angeles, I was treated to a simple diagram of electrical wiring superimposed over everything I saw. I was surprised at how quickly and easily I adjusted to the diagram's presence. It was neither an annoyance nor a distraction, but just another object – albeit virtual – in my field of view.

To all outward appearances, the VRD is rather unassuming – a pair of high-tech eyeglasses attached to what looks like a cell phone worn on a belt. Inside the cell-phone-shaped box, which is actually a computer, are photon generators: small laser diodes or light-emitting diodes that produce light signals. (Diodes function as detectors or emitters of infrared or visible light when charged with an electric current.) Feed an image into the computer and a scan-converter sends the signals to the glasses through a fibre-optic cable. Microscanners then process these signals to produce images that are cast onto the retina. The VRD displays a picture that doesn't block the user's view but is superimposed over the standard field of vision. The result is a complete picture – of a map, a sign or a face – that appears to be floating directly in front of the viewer at about arm's length.

The VRD uses a beam of low-power laser light – about the equivalent in intensity of 'going outside in daylight', according to Furness – to scan images onto the retina. The images are built up in a raster, not unlike the way conventional televisions and computer monitors work. The scanner rapidly moves in a horizontal direction across the retina depositing a row of pixels as it goes, while a second scanner does the same in a vertical direction. The image is constructed like a pointillist painting daubed directly onto the retina, the sequences of individual dots merging to form a realistic, high-resolution, three-dimensional picture.

This experimental version of the VRD is semi-portable, but not yet easily wearable. Eventually, Furness predicts, the electronics will be reduced to a unit about the size of a pack of cigarettes that can be comfortably worn on a belt. The display device itself will be about the size of a thimble. The first commercial products are relatively bulky, though. There are two types of VRD currently in use: a bench-mounted unit that displays a full-colour image and a portable unit that displays only black and white. Microvision, the Washington-state-based company that is commercialising the VRD, is developing smaller displays to be incorporated into eyeglasses, goggles and helmets. The firm also intends to integrate the device into cellular phones and pagers, allowing users to tap into business networks or the Internet to view e-mails, Web pages, faxes and files as if on a full-sized desktop monitor. 'One of the most important interfaces of the future will be vision,' says Furness. 'An early application of the VRD will be in conventional displays for cell phones – just look into the cell phone screen and the VRD will project images into your eye.'

While the images conjured up by the VRD may be virtual, the potential benefits are very real. 'After we finished the first demonstration in 1993,' Furness recalls, 'a person came into the lab, looked into the display and said, "I can see this image perfectly with my blind eye." We were blown away, and started investigating.'

Furness's group learned that the person whose sight was restored had been in an automobile accident several years before. His injured eye was obscured by scar tissue that prevented light from reaching the retina, leaving him effectively blind in that eye; but the retina itself was still intact – and the VRD image flowed right through tiny gaps in the scar tissue to strike it. This serendipitous discovery gave Furness his first hint that the VRD could help people with some visual impairments, such as scar tissue and cataracts, see again, possibilities that are now under investigation in his lab.

In addition to restoring sight to the blind, the VRD has applications for those with standard vision too. The 'augmented vision display', which Furness describes as images of 'the normal outside world with the virtual world superimposed on top', would enable architects and contractors to construct buildings around virtual images of the planned structure. Using the VRD as an elaborate paint-by-numbers set, construction workers could assemble a building simply by filling in the virtual image with the actual materials. 'You will see a virtual girder,' Furness explains, 'and put the real girder in the same place. We can provide a kind of hypervision, enabling people to see things they would never see otherwise.'

These kinds of display are already in use at the Wallace-Kettering Neuroscience Institute in Cleveland, Ohio. Instead of seeing virtual girders, neurosurgeons wear a head-mounted VRD to view brain scans superimposed directly onto the patient's head during surgery. This internal view of the brain helps guide the surgeon's hands during delicate procedures. Furness and his colleagues at the HIT lab are also developing an interactive VRD for the US Navy that would allow pilots to easily view a wide range of flight data – including virtual scenes projected onto the actual view from the cockpit – and activate controls merely by looking at them.

Computer games are another area in which VRD technologies are likely to prove popular. What ardent player wouldn't welcome the chance to liberate their virtual worlds from the confines of the computer monitor and have their favourite heroes and villains appear before their very eyes? As graphic image processors and VRDs become increasingly powerful, games will reach unprecedented levels of visual realism. Furness even predicts the development of 'Watchmans', hand-held devices like Sony's Walkman capable of displaying images on the retina that appear as large as an I-Max movie theatre screen. 'You can be sitting on a bus but you'll feel as if you're in the theatre and the screen surrounds you,' Furness says.

This type of technology is already finding its way into consumer products. 'Videos in the desert! Formula 1 in the garden! Cinema in bed! Don't let your TV set dictate where you are entertained!'

the Eye-Trek website exhorts. The Eye-Trek TV system, made by Olympus, is a personal video screen and headphones ensconced inside a pair of wrap-around sunglasses. When connected to a video source – a television, video player or digital video disc – the device projects the television programme or film onto the glasses. The company claims that viewing the image is like sitting two metres away from a 1.30-metre-wide screen. Eye-Trek enables users to watch TV any-where within a 50-metre radius of the source. The device does place some strain on the eyes during focusing, however. While Olympus says this is not a problem for adults, it recommends that Eye-Trek not be used by anyone under sixteen years of age. Since Eye-Trek is portable, it could well be difficult to patrol who uses it, especially since it is likely to appeal to teenagers, the same age group for whom it is not recommended; and there's another potential glitch that could limit the appeal of these immersive virtual technologies, regardless of the user's age: cybersickness.

Cybersickness is the feeling of nausea and disorientation that can occur when the sensory cues coming from a virtual image conflict with those coming from objects in the actual environment. Furness calls cybersickness 'a potential showstopper' for the development of VRD-like devices, so he and his collaborators are trying to find ways to prevent the condition. The biggest problems occur when there is a conflict between what the eyes see and what the vestibular system, the mechanisms of the inner ear that maintain balance, tell the brain. When the virtual scene moves but the body does not, the brain is confronted with a conflict between two sensory inputs and doesn't know which one to believe. The most promising technique developed to date is to project a grid pattern over virtual images that actually moves in sync with the body.

There may, however, be another danger inherent in devices like the VRD: will the virtual worlds they make possible become so compelling that users lose interest in the actual world around them? 'VRD technology is no more a problem than television,' says Furness. 'But the more photorealistic the virtual world becomes, the more difficult it is to tell it apart from the real world. The VRD makes virtual reality better, so good in fact that it's possible to get lost in it. It can be engaging enough so that it becomes easier to get addicted.'

Cyber-addiction may not be the most worrying or even the main side effect of the VRD, however. After all, couch potatoes and bookworms – people who are addicted to television and books, respectively – have been with us for a long time now and civilisation as we know it has not yet ended. The bigger threat might be that VRD worlds become so dazzling, detailed and compelling that they kill the imagination, especially among young people who are likely to be most drawn to the technology.

Furness is aware of, and concerned about, the risk. 'Virtual worlds are happening all around us anyway,' he says. 'Their development is inevitable. The question is: how can we use them for good rather than just as mind candy? If all we can watch on the VRD is daytime talk shows, it won't be very satisfying. It's up to the content providers – poets, scriptwriters, film-makers – to make it a worthwhile experience.'

The irony of this dilemma is that imagination itself is a virtual technology, too. Every time a poet writes a poem, a painter paints a painting or a composer composes a piece of music, a virtual world is created; but appreciating works of art is just as much an act of the imagination – and therefore a virtual act – as creating them. Writing a poem and reading a poem are both creative acts, requiring strenuous imaginative exertions. Yet each is incomplete without the other. Great works of art invariably leave empty spaces that can only be filled and inhabited by the viewer's imagination.

So the danger is not that devices like the VRD are by their very nature imagination-killers; they are not. On the contrary, the new possibilities the technology opens up could well be a spur to further creativity, perhaps even leading to the evolution of entirely new art forms, a possibility discussed in the next chapter. The risk, as Furness points out, is that virtual worlds become so immersive that people start to drown in them; that they stop doing the hard work of imagining themselves and let the VRD do it for them; that the sacred spaces of the imagination gradually get filled up with nothing more than idle daydreams.

The virtual worlds conjured up by the VRD might be seductive, but in many ways they are no less 'true to life' than the things we see with the naked eye. We don't perceive the world directly through any of our senses. The sensory information that finally arrives at our brain is a translation at best, in many instances a construction, and in some cases an outright fabrication. For no sense is this truer than for vision.

To start at the most basic level: while light waves enter the eye, the information registered in the brain comes from electrical and chemical impulses. So we don't perceive the light directly, but only the optic nerve's interpretation of that light. Also, the image that forms on the retina is flat, two-dimensional; but what we see is in three dimensions. The brain itself builds in the depths and surfaces; and we've already seen how the visual system selectively edits the information that reaches the brain. Information that doesn't change tends to get ignored.

I was confronted with this fact recently when a building near my office in central London was torn down. I used to walk past that building five or six times a day, so often that I no longer even noticed it. But after it was torn down, a whole new vista on that street corner was suddenly opened up to me. I noticed things I had long ago stopped seeing: the small busts carved into the façade of the building across the street, the alley that ran along behind what was now a construction site, how steeply the land sloped down toward the Thames. It was as if I looked at the neighbourhood with new eyes.

So, as we ponder the influence of virtual technologies on our lives, it's important to remember that our senses are virtual too. Only a small fraction of the world out there ever gets inside us, and what does enter is actively shaped and created by our senses and our minds. As American psychologist William James wrote, 'Whilst part of what we perceive comes through our senses from the object before us, another part (and it may be the larger part) always comes out of our own mind.' What you see is not so much what you get as what you make.

THE BIONIC MANN

Steve Mann, professor of electronic engineering and computing at the University of Toronto, has a mind to shape what he sees — and he's invented the technology to do it. 'I have melded technology with my person,' he wrote in MIT's *Technology Review* magazine. 'Every morning I decide how I will see the world that day. Sometimes I give myself eyes in the back of my head. Other days I add a sixth sense, such as the ability to feel objects at a distance.'

Mann gets his extrasensory perception from the WearComp, a wearable computer he began developing more than twenty years ago in Canada. The current WearComp is a wireless wardrobe of wearable gadgets – miniature video cameras and screens, microphones, portable microprocessors, finger-sized joysticks and an array of sensors that monitor the user's vital signs as well as the external environment – that augments and extends his senses.

Ever wish, like Mann, that you had eyes in the back of your head? Try on the VibraVest for size and you will. The VibraVest, a computational tank top worn underneath ordinary clothing, is a portable radar system originally invented to assist the blind. Flexible electronic circuits are sewn into the VibraVest fabric, which is also woven with radio transceivers that vibrate to alert the wearer to the approach of distant objects. When activated, the VibraVest enables the wearer to feel objects pressing against the body even though they are actually far away from the body in space.

Say you are cycling through busy London traffic and a black cab comes speeding up from behind. The VibraVest detects this and translates the information into a steady pressure on your back. As the taxi starts to get too close for comfort, the pressure becomes more and more intense. Since you feel what the VibraVest 'sees' Mann calls the result 'synthetic synaesthesia', a mixing of the senses brought about through electronics. Devices like the the VibraVest are compact, lightweight, unobtrusive and in many cases sewn into conventional garments – even underwear. Mann's devices are based on 'humanistic intelligence', a design approach to wearable computing in which the user's mind and body are part of the information-processing pipeline.

The Xybernaut corporation in Fairfax, Virginia, has been making wearable computers based on Mann's designs for the past ten years. The company's Mobile Assistant IV (MA IV) does everything a desktop PC can do – access data, send e-mails and connect to the Web – but does it on the fly via a portable microprocessor, a head-mounted display and speech recognition technology. The MA IV is used primarily for industrial tasks that require both mobility and manual dexterity, like aircraft maintenance. It takes roughly two tons of technical manuals to keep a Boeing 777 running. Rather than lug around the necessary tomes, Boeing engineers consult the required information via the MA IV's head-mounted display, calling up the appropriate pages by voice command, thus keeping their hands free for repairs.

Wearables won't long be confined to the workshop alone, though; they're already turning up on the catwalk, too. The Dutch electronics firm Philips, for example, has developed a range of 'smart fabrics' that look like ordinary children's clothing but contain a global positioning system and video camera so that parents can check up on their kids remotely. Those kids may be wearing digital denim jackets with audio speakers woven into the collars – simply flip up the collar around the ears to listen your favourite tunes – so they may not hear their parents calling via the mobile phone embedded in the pocket. Golfers can dress for success on the green with intelligent apparel that analyses and helps them improve their swings. And aficionados of extreme sports can rest assured that their gear will automatically alert emergency services should they have a nasty spill while snowboarding in the Himalayas. 'We are entering a pivotal era,' Mann says, 'in which computational technology will become part of our everyday lives in a much more immediate and intimate way.'

Mann has become very intimate with his WearComp8, the latest generation of the device. The WearComp8 is a complete multimedia computer with cameras, microphones and earphones built into an ordinary pair of sunglasses. Mann wears the system up to sixteen hours a day, sometimes even sleeping with it, and says it's suitable for any situation 'other than bathing or during heavy rainfall'. Mann has become so attached to his WearComp that he often feels a bit uncomfortable when he takes it off, as if he is missing some essential article of clothing. 'People find me peculiar,' he says. 'They think it's odd that I spend most of my waking hours wearing eight or nine Internet-connected computers sewn into my clothing and that I wear opaque wrap-around glasses day and night, inside and outdoors. But the system is situated in my own personal space. I regard it as part of me, and others do as well.'

Though previous versions of the head-mounted display were heavy and cumbersome, the current one is practically indistinguishable from conventional shades. Its capabilities are anything but conventional, though. Ensconced inside the frame on one of the lenses is a computer screen that can display Web pages, e-mail messages, family snapshots, city maps and much else besides. Video cameras concealed in the glasses can be oriented to have the same field of view as the wearer or can afford different points of view, such as a rear-facing camera that monitors the scene directly behind the user. The WearComp is similar to Tom Furness's virtual retinal display, except that Mann sees images on a screen and not projected directly onto his retina. Instead of using the WearComp simply to view what's out there, Mann uses it to actively shape what he sees – a third eye that filters or augments all he surveys without obstructing his normal view – unless he wants it to.

Fed up with the surfeit of advertisements in the world? Program the WearComp to recognise billboards and posters and it will filter them out of your field of view. Should you want to enhance rather than curtail your purview, the WearComp can do that too through the EyeTap.

Mann has come up with a way for WearComp-users to annotate and embellish the world by means of 'homographic modelling', a technology that enables people to stick virtual Post-it notes on physical objects. Want to remind your spouse that you've got theatre tickets for tonight? Simply send a virtual note and attach it to his office door. When he arrives at work, the EyeTap recognises the office door, detects the virtual note attached to it and calls up the message so that it appears to be actually pasted on the door. Since the message is intended for his eyes only, only his EyeTap can see it.

The WearComp-user can also allow another person to share or alter their perspective on the world. The WearTel™ phone uses EyeTap technology to allow individuals to see each other's points of view. 'The WearTel phone let's someone be you rather than just see you,' Mann says. Imagine strolling under the garish neon lights of Tokyo's Akihabara district sporting the WearTel. A friend or family member back in London could access the device's video camera through the Internet and enjoy the view from over your shoulder while you describe your impressions. Less exotic applications might include providing remote guidance during a shopping expedition or giving a motorist directions as he drives through an unfamiliar neighbourhood. More adventurous WearTel users could cede control of their vision to someone else, allowing that person to alter their perceptions by introducing visual distortions to the scene or adding virtual images.

Mann is convinced that, unlike many other forms of personal technology, the WearComp will empower rather than distract the user. 'The WearComp is built on the assumption that computing is not the primary focus of attention,' he explains. 'It doesn't cut you off from the outside world like a virtual-reality game. You can attend to other matters while using the apparatus.'

The WearComp is always active while worn, operating in the background and only grabbing your attention when you want it to – or when it has something unusual or important to tell you – which is, in effect, exactly the way our biological senses work. 'It is misleading to think of the wearer and the computer as separate entities,' Mann says. 'Instead, it is preferable to regard the computer as a second brain and its sensory modalities as additional senses, which are inextricably intertwined with the wearer's own biology.' The result will be what Mann calls 'computationally mediated reality', a perspective on the world that is not given but is actively shaped and created by a person's wearable computers. Ultimately, Mann believes, we will arrive at a 'completely computer-mediated world in which all aspects of life will be online and connected'.

FACE TO FACE

Mann is one of only a handful of people currently using computer technology to change their perspective on the world; but these days, walk down almost any city street and you can be sure that at least half a dozen closed-circuit video systems are watching you. Soon computerised video networks could be noticing more than just who happens to stroll by.

Software called BlueEyes, developed at IBM's Almaden Research Center in San Jose, California, is used by retailers in surveillance systems to monitor and interpret customer behaviour. Video cameras record shoppers as they browse the aisles. The BlueEyes software studies customers' eye movements and facial expressions. For example, your pupils dilate when you're interested in something in order to take in more visual information. If BlueEyes notes very few dilated pupils in front of a particular display, the store manager might conclude that that advertising campaign is a dud. Conversely, should BlueEyes register lots of wide-open pupils and smiling faces, the manager might well try similar promotions elsewhere in the store. The potential for abuse is clear, and technology like BlueEyes is meeting resistance from civil liberties groups.

Similar technology, this time developed at the Salk Institute in La Jolla, California, watches the same telltale facial signs, not to discern shopping patterns but to determine if you're lying.

Deception comes naturally to all living things. Birds do it by feigning injury to lead hungry predators away from nesting young. Spider crabs do it by disguise: adorning themselves with strips of kelp and other debris, they pretend to be something they are not, and so escape their enemies. Nature amply rewards successful deceivers by allowing them to survive long enough to mate and reproduce.

So it may come as no surprise to learn that human beings – who, according to psychologist Gerald Jellison of the University of Southern California, are lied to about 200 times a day, roughly one untruth every five minutes – often deceive for exactly the same reasons: to save their own skins or to get something they can't get by other means.

Knowing how to catch deceit can, however, be just as important a survival skill as knowing how to tell a lie and get away with it. A person able to spot falsehood quickly is unlikely to be swindled by an unscrupulous business associate or hoodwinked by a devious spouse. Luckily, nature provides more than enough clues to trap liars in their own tangled webs – if you know where to look. Researchers at the Salk Institute have programmed a computer to get at the truth by analysing the same physical cues available to the naked eye.

No lie detector, technological or biological, detects lies as such; it merely detects the physical cues of emotions, which may or may not correspond to what a person says. Polygraphs, for instance, measure respiration, heart rate and skin conductivity, the electrical current of the skin, which normally increases with stress or excitement. These signals tend to leap when people are nervous, as they usually are when lying. So a sudden surge in skin conductivity and an outbreak of perspiration could indicate nervousness – about getting caught, perhaps? – which might, in turn, suggest that someone is being economical with the truth. On the other hand, it might also mean that the person is simply too hot, which is one reason polygraph tests are inadmissible in court.

So instead of focusing on potentially misleading physical cues, the Salk Institute's lie detector homes in on clues to deceit that are difficult to fake or conceal; and those clues are written all over the face.

Because the musculature of the face is directly connected to the areas of the brain that process emotion, the countenance can be a window to the soul. Neurological studies even suggest that genuine emotions travel different pathways through the brain from insincere ones. If a patient paralysed by a stroke on one side of the face, for example, is asked to smile deliberately, only the mobile side of the mouth is raised; but tell that same person a funny joke, and the patient breaks into a full and spontaneous smile.

Very few people – most notably, actors and politicians – are able consciously to control all of their facial expressions. Lies can often be caught when the liar's true feelings briefly leak through the mask of deception.

One of the most difficult facial expressions to fake – or conceal, if it is genuinely felt – is sadness. When someone is truly sad, the forehead wrinkles with grief and the inner corners of the eyebrows are pulled up. Paul Ekman, professor of psychology at the University of California, San Francisco, who has been studying facial expressions for the past thirty years, found that fewer than 15 per cent of the people he tested were able to produce this eyebrow movement voluntarily. By contrast, the lowering of the eyebrows associated with an angry scowl can be replicated at will by almost everybody. 'If someone claims they are sad and the inner corners of their eyebrows don't go up,' Ekman says, 'the sadness is probably false.'

The smile, on the other hand, is one of the easiest facial expressions to counterfeit. It takes just two muscles – the *zygomaticus major* muscles that extend from the cheekbones to the corners of the lips – to produce a grin. But there's a catch. A genuine smile affects not only the corners of the lips but also the *orbicularis oculi*, the muscle around the eye that produces the distinctive 'crow's-feet' associated with people who laugh a lot. A counterfeit grin can be unmasked if the lip corners go up, the eyes crinkle but the inner corners of the eyebrows are not lowered, a movement controlled by the *orbicularis oculi* that is difficult to fake. The absence of lowered eyebrows is one reason why false smiles make a person look constipated rather than happy.

In the 1970s, a team of psychologists led by Ekman classified all the muscle movements – ranging from the thin, taut lips of fury to the arched eyebrows of surprise – that underlie the complete repertoire of human facial expressions. In addition to the nervous tics and jitters that can give liars away, Ekman discovered that fibbers often allow the truth to slip through in brief, unguarded facial expressions. Lasting no more than a quarter of a second, these fleeting glimpses of a person's true emotional state – or 'microexpressions', as Ekman calls them – are reliable guides to veracity.

A computer program developed by the Salk-led team has been trained to catch these microexpressions and detect incongruous or inappropriate facial expressions that might indicate deception. The program works by comparing video images of faces to sixty different templates, each of which is focused on a specific muscle movement in a specific facial region. If a subject professes sadness, for example, but the inner corners of their eyebrows don't go up, the computer notes this and, after compiling and comparing the data from all sixty regions, will assess how that person's expressions match the muscle movements that correlate with genuine sadness.

In one study, the program was trained to recognise six of forty-six individual muscle actions described by Ekman. For all six actions, it outperformed human non-experts and performed as well as highly trained human experts in lie detection. The Salk investigators hope to expand the program's repertoire of expressions and develop applications for law-enforcement officials and mental-health professionals.

Hiroshi Kobayashi and Fumio Hara, professors of mechanical engineering at the Science University of Tokyo, have found a different application for machine vision: a robot that detects human facial expressions and responds with facial expressions of its own. The face robot, which resembles a young Japanese woman with black hair, can express surprise, fear, disgust, anger, happiness and sadness through eighteen actuators installed around its eyes, nose, chin and mouth.

When I visited Kobayashi's lab, the robot was disassembled for repairs. Its face, made of flesh-like silicon rubber, lay crumpled in a flaccid heap on a table like a discarded Halloween mask. The hardware looked eerily like the scene from the 1973 sci-fi film *Westworld*, in which the murderous gunslinging robot played by Yul Brynner has its face removed in the laboratory to reveal a tangle of wires and circuits. The head of Kobayashi's face robot is filled with circuitry and clusters of shape-memory alloys, ultrathin wires that contract when electrical current is sent through them. The wires sprouted limply from the faceless skull, like a marionette's strings awaiting the hand of a puppeteer.

This is, in a way, exactly what they are. The robot recognises the facial expressions of people through a CCD camera installed in its left eye. The CCD detects a face, converts images of the facial expression into electronic signals and sends the information to a processing unit. The processing unit identifies the expression and sends instructions back to the robot to mimic that expression. If you smile at the robot, it smiles back; if you frown, it frowns too.

The computer manages the robot's expressions by sending electrical charges through the appropriate shape-memory alloys. When a charge is applied to the wire it contracts, pulling the robot's flesh into different expressions. With the right application of electrical current, the robot can raise and lower its eyebrows, wrinkle its nose, pull back the corners of its lips and open its mouth. When the electrical charge stops, the wires return to their normal shapes and the robot's face relaxes. Kobayashi and Hara followed Ekman's muscle classification system in deciding where to place the wires.

Though the robot itself was not operational during my visit, I was able to see a video of the face in action, and it was quite unsettling. Even when expressionless, the robot has a strange, unnatural look. I felt a little bit as if I had sneaked into a wax museum after closing time. Because the face has only a handful of control points, the expressions are still relatively crude. The smile, for example, is more of a grimace than a grin, giving the robot a pained expression. The nose wrinkles quite effectively, though. Kobayashi is the first to admit the difficulty of making the face appear natural and is working to add more control points to increase the robot's expressiveness.

It is a challenge he will have to overcome if he wants these kinds of devices to become popular as a new kind of computer interface. Kobayashi imagines face robots as user-friendly ways for the elderly to access the Internet or as companions for older people who live alone. 'An elderly person could have the face of his or her son or daughter as an interface,' Kobayashi says. 'The face could talk to the user, and older people who might require observation for health reasons could be watched through the eyes of a loved one.'

Though it may sound eerie, computers that have a friendly face and are good listeners are being put to some surprising uses, as we'll see in the next chapter, on 'Hearing'.

THREE

HEARING

The Sounds of Science

'My God! It talks!' EMPEROR PEDRO OF BRAZIL, AFTER HEARING A DEMONSTRATION OF THE TELEPHONE IN 1876

Baldi is an excellent listener. For many of the students at the Tucker Maxon Oral School in Portland, Oregon, Baldi – an animated conversational agent, developed at the University of California, Santa Cruz – *is* a person. Baldi teaches the profoundly deaf children at the Tucker Maxon school to talk by working with them to improve their reading, spelling and speech-production skills.

Because they can't hear the daily chatter around them, deaf children don't pick up spoken language naturally like children with normal hearing. Baldi, an animation of a hairless human head, helps deaf youngsters between the ages of six and twelve overcome this deficit through a kind of talking cure: the students talk while Baldi listens and demonstrates the correct way to produce various sounds. 'The kids love it,' says Pamela Connors, a linguist who works with Baldi and the children. 'As soon as the technology was brought in, everyone started calling it Baldi. The students treat him as part of the class, and say things like, "He's a good teacher," and "He understands me." '

Most of the children at Tucker Maxon have cochlear implants, devices inserted in the inner ear that electronically deliver signals to the auditory nerve. Some have hearing aids. The implants restore enough of the children's hearing to enable them to perceive a limited range of sounds, including conversational speech. Baldi is used at the school as part of a speech-technology package that includes a synthetic speech generator, a text-to-speech system, a speech-recognition system and an animated head.

To engage Baldi in conversation a student simply speaks into a microphone in response to an auditory prompt from Baldi. Baldi might start by asking, 'How are you?' and the subsequent conversation will differ depending on whether the child says 'Fine, thanks' or 'Not so great, really' in reply. In either case, Baldi listens to the student and runs the sounds through a speech-recognition system. This system notes the mispronunciations and then launches a dialogue designed to correct the child's mistakes. Baldi does this through vision as much as through hearing.

When engaged in conversation, whether it be idle gossip, heated debate or philosophical speculation, we glean as much information from the other person's face as we do from his or her actual words. Children use these visual cues – the shape of the lips, the positions of the tongue and teeth – to associate sounds with particular movements of the mouth. The hearing-impaired can't compare the sound with the movement, though, so these visual cues become even more important, since they are their only clues about how to produce oral speech.

Baldi teaches students to produce more accurate utterances by showing them how it's done. It shows them how to produce a 'sh' sound by puckering up its lips; it shows them how to produce a 'th' sound by sticking its tongue between its teeth; it can even become transparent to show them how the tongue hits the roof of the mouth when producing a 'l' sound. Baldi can also change the orientation of its face while speaking so that it can be viewed from different per-spectives. For a more private tête-à-tête, kids can plug their cochlear implants or hearing-aid processors directly into the computer's audio output (and thus into Baldi itself) in order to avoid the distractions of classroom noise. Throughout the dialogue, Baldi explains to the student exactly what it's doing with its lips, mouth and tongue.

After completing a lesson with Baldi, the student can record him- or herself saying the words. If they like what they hear, they can save the recording in a personal directory so that the therapist or teacher can listen to it later. Pamela Connors says students, parents and teachers alike have grown quite fond of Baldi, and recognise the technology's contribution to helping the children speak more accur-ately and expressively: 'The teachers want to immerse the students in oral language. They use Baldi as a classroom supplement because his visual speech is quite accurate – like a human, not a cartoon – and he is consistent and patient. The students like him because they can both hear and lip-read him, as well as change his volume and have him repeat a question or word as often as they like.'

While Baldi is teaching students at the Tucker Maxon school to talk, the fact that they can hear at all is due to cochlear implants, the single most successful application of advanced electronics for sensory repair to date.

Deafness usually occurs as a result of trauma, genetic or otherwise, to the roughly 3500 hair cells found in the cochlea, a bony structure shaped like a snail's shell that's located deep within the inner ear. These cells do not regenerate or replace themselves when they are damaged or die, so permanent deafness results.

Like geraniums in a window box, the hair cells are planted in tidy rows along the basilar membrane, a patch of tissue at the base of the cochlea. The basilar membrane serves as a sounding board for the three tiny bones of the middle ear – the malleus, the incus and the stapes – that leap into action in response to the mechanical vibrations of sound. When sound waves cascade, surge or crash through the spirals of the inner ear, the stapes (also known as the stirrup because of its shape) starts tapping furiously at the basilar membrane like a telegraph operator sending out urgent Morse code.

This percussion sends out tremors along the basilar membrane, which in turn sets the hair cells bouncing, jiggling and generally swaying to the stapes' beat. Each movement of a hair cell fires off a burst of electrical signals through the auditory nerves to the cochlea, which forwards the signals to the hearing centres of the brain where they are converted into a recognisable sound – a cymbal crash, for example, or the faint exhalations of a sleeping child.

While conventional hearing aids simply amplify sound – in effect, giving hair cells a bigger jolt by pumping up the volume – the cochlear implant uses a microphone and processing unit to bypass the damaged hair cells altogether and deliver electrical signals directly to the auditory nerves, in much the same way as Claude Veraart's visual prosthesis bypasses dysfunctional photoreceptor cells to reach the optic nerve. The implant itself consists of an external microphone and a small speech processor, which rests behind the ear. A receiver is implanted deep inside the ear and is connected to the cochlea by eight electrodes.

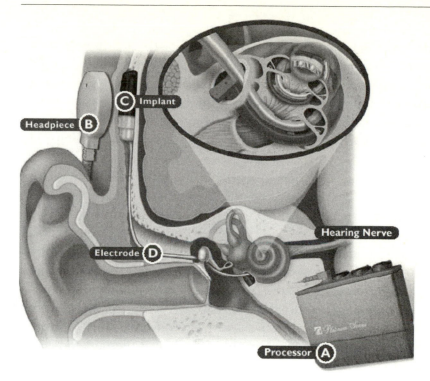

Figure 4. The Clarion™ cochlear implant.

The cochlear implant 'hears' when sounds are picked up by the microphone and sent to the processor, which converts them into an electrical signal. This signal is transmitted via radio waves to the implanted receiver, which sends a rapid series of pulses to the electrodes in the cochlea. From there biology takes over, and the cochlea stimulates the auditory nerve to conduct impulses to the brain's hearing centre. Thus the cochlear implant is not merely giving ailing hair cells a boost; it's replicating electronically what the human ear does naturally.

The cochlear implant is installed under the skin just above and behind the ear. None of the implanted electronics is visible once the incision has healed. Users don't hear in the normal sense until the speech processor is programmed, and even then many people report that the first sound sensations they perceive are distorted and confusing. It can take up to several months before a user adjusts to the device.

At MIT Rahul Sarpeshkar, whom we met in the chapter on 'Sight', is working to make the cochlear implant even more like the biological system it emulates, and thereby help improve its performance. Sarpeshkar cites the human ear's ability to perform extremely sophisticated computations in an impressively small amount of space and with an impressively small amount of power – feats that still can't be matched in real time by today's powerful computers. He estimates that the power consumption of the ear is so small that it could run on an AA battery for fifteen years. 'People often underestimate what a marvellous instrument the ear is,' he says.

Sarpeshkar has built a silicon cochlea chip that models the biophysics of the human cochlea in some detail. This chip replicates the propagation of mechanical displacement waves on the basilar membrane and fluid pressure waves in the cochlea, which together generate the impulses that the brain interprets as sound.

In the biological cochlea these waves travel down the length of the cochlea, set in motion by the piston-like movement of the stapes. The waves instigate the movements of the basilar membrane, which triggers the motions of the inner hair cells, which, in turn, generate electrical signals in the auditory nerve fibres. High-frequency waves caused by high-frequency sounds travel only short distances in the cochlea before they are cut off, while low-frequency waves caused by low-frequency sounds travel longer distances. Auditory nerve fibres near the base of the cochlea respond to higher-frequency sounds and those near the apex respond to lower-frequency sounds.

Sarpeshkar says that one of the key advantages of this travelling-wave architecture is that 'it can produce enormous amounts of amplification by causing the wave to grow gradually but steadily as it propagates down the cochlea'. Such amplification would be very hard to produce if it were all done in one place, yet without it we would be unable to detect many subtle but complex sounds, from an orchestra tuning up before a performance to the last, exhausted gasp of a deflated balloon.

The cochlea's outer hair cells amplify the motion of the basilar membrane, so that a small change in the gain provided by the hair cells eventually makes a very large change in the overall gain of the sound. 'It's like compound interest,' Sarpeshkar says. 'A small change in the interest rate can accumulate over time to cause a large change in the overall growth of an investment, just as a small change in the gain provided by a cascade of outer hair cells can accumulate over space and cause a large change in the overall growth of the cochlear wave.'

Sarpeshkar's chip mimics the travelling-wave architecture of the cochlea and models the motion of the basilar membrane as well as the amplification provided by the inner and outer hair cells. The chip manages to do all this while consuming just 0.5 mW of power, sufficiently low that it could run on an AA battery for a year and yet still operate in real time. Sarpeshkar says that low power consumption and real-time operation are crucial for such implants, since replicating the computations of the biological cochlea on a slow and power-hungry computer would make the device impractical.

Sarpeshkar is currently putting silicon-cochlea chips to work inside his own version of the cochlear implants. Nevertheless, he admits that his silicon version has a long way to go before it can even begin to match the performance of the human ear. 'Biological systems are so complex and so superbly designed that the man-made stuff is far behind,' he says. 'The human cochlea has 3500 hair cells and 30,000 nerve fibres; the electronic version has just eight electrodes. Yet the brain can still make sense of the signals. That shows how good the brain is, not how good these artificial cochlea are. We still have a very long way to go before learning how to compute like biology.'

LOOK WHO'S TALKING NOW

Biology, though of an alien kind, is the first thing that comes to mind when looking at the three basketball-sized eyeballs gently bobbing and jiggling on a table in Michio Okada's lab at the Advanced Telecommunications Research Institute (ATR) near Kyoto, Japan. The pastel-coloured blobs – light blue, purple and yellow – look like dollops of freshly squeezed toothpaste, each with a little swirl on top. As they bounce around on the table top the eyeballs babble like infants, emitting a low murmur of nonsense syllables. Meet Muu, embodied interfaces for human–computer interaction.

According to Okada, Muu (which means 'eyeball' in Chinese) are 'artefacts that can establish social bonds with human beings. A lot of people, especially the elderly and children, need easier ways to interact with computers. In order to draw out our natural communication skills, a different type of interface is required.'

Muu are certainly different. These alternately comical and disconcerting creatures can see through their one large eye, which harbours a video camera in the pupil so that they can locate and attend to their conversational partner. They also have tactile sensors and respond with yet more excited wiggling when someone touches their soft exterior. And they can, of course, hear and even initiate conversations, albeit chats limited to abstract emotive expressions like 'That's too bad' or 'That sounds good!'

The point, Okada claims, is that people like a good natter, even when they have nothing much to say, as is shown by the popularity of short messaging services – brief text missives sent over mobile phones – among teenagers. Through Muu, Okada is trying to use this chatter as a more intuitive way for users to access computers and the Internet. Conversational interfaces like Muu are made up of software that comprehends written or spoken human language and responds with text or synthesised speech. Speech recognition and speech generation, as this suite of technologies is called, lends computers ears and a voice so that we can talk to them in much the same way as we talk to one another.

Okada employs a traditional Japanese children's language game to explore how users respond to Muu. The game, called *siritori*, involves two or more players. The first player says a word, and the next player has to use the last syllable of that word as the first syllable of the next word. For example, the first player might say the word 'open' and the second player could answer with 'enter'. The game then continues along the same lines; in this case, the next word might be 'terror'.

Okada says the kids who have played *siritori* with Muu loved it, and found it a much more natural way to interact with a computer than clicking on icons. More important, he believes, is the fact that the children were able to build a relationship with the device, a key factor in the development of more intuitive interfaces. He's convinced that Muu could become a helpful addition to the classroom to assist kids in learning languages, for example.

While this kind of interaction may be great for children, will adults ever want to access the Internet through the likes of Muu?

Karl L. von Wendt thinks so. 'The Internet is often compared to the invention of the printing press, but I believe it is far more important than that,' says a smiling Von Wendt, CEO of Kiwilogic, a Hamburg-based firm specialising in artificially intelligent virtual characters that can understand and respond to written language. Kiwilogic's virtual agents interact with users in conversational mode – just type in an inquiry and the agent provides an answer based on an analysis of keywords in the question. Von Wendt calls the agents 'lingubots'.

'Aren't you sick of monotonous virtual creatures that are nice to look at but can't even answer a simple question?' Von Wendt says. 'With lingubot technology you can bring your mascots and symbols to life – your customers will suddenly be able to speak to them!'

'Do people really want to speak with their computers?' I ask.

'The idea behind a lingubot is to make the Net more human via the medium of natural dialogues,' Von Wendt says with a grin. 'So far lingubots have only been used on Web pages, but it is also possible to install a bot offline. Possible areas of application would be computer games, intranets and all other situations in which humans interact with computers.'

'What about you?'

'Okay, you are talking to a software application here. I am a lingubot, and I am an example of what we develop here at Kiwilogic.'

This conversation took place not with the biological Karl L. von Wendt but with his virtual doppelgänger, an intelligent agent that Kiwilogic calls the 'first online virtual CEO in the world'. To chat with the virtual Von Wendt I logged on to the Kiwilogic site and typed in my questions. Von Wendt responded with written text, accompanied by a series of still images of him displaying appropriate facial expressions – frowning, grinning, giving the thumbs-up sign.

The virtual Von Wendt isn't exactly a scintillating conversational partner – he has a rather limited knowledge base and tends to repeat himself – but his online patter isn't just strictly business. When I asked what his favourite food was, for example, he grinned broadly and said, 'I love everything that is sweet – unfortunately.' And when I asked if he believed in God, he got all serious-looking and said, 'That is one thing we lingubots have in common with humans: we think we know everything, but we don't.'

The flesh and blood Von Wendt may not know everything either, but one thing he is sure of is that speech recognition will transform the way people use their computers and access the Web. 'There's nothing more annoying than talking to a stupid computer system that keeps telling you to "Press star now" or "Push 1, 2 or 3",' he says. 'We are making the Web talk so that it understands natural language and can engage in real dialogue. Soon we will all be talking to our machines, but the machines will be intelligent.'

Why would we want to talk to our machines? One reason is convenience. The Web is becoming increasingly important as a primary source of information for everything from news, sports and weather forecasts to stock quotes, theatre listings and airline reservations. But as any inveterate surfer will tell you, trolling the Web for information can be a click-and-miss affair, a tedious, time-consuming and frustrating process that often leads nowhere. The Web experience would be a lot more efficient and pleasant if the computer could understand what you were looking for and find it for you.

Systems with this ability already exist. In the US, Tellme is a free phone-based service through which users can receive information about restaurants, taxis or the stock market. Just call the Tellme number and say a keyword, such as 'movies', into the receiver. The computer listens to your query, prompts you with requests for more information, such as the name of your local cinema, and recites a list of films and showing times at the desired theatre. You can interrupt at any time or issue commands such as 'go back' or 'next' to move quickly through the lists. Should you want to reserve tickets, Tellme will put you through to the box office.

Wildfire Communications offers a similar but more sophisticated phone-based service. Wildfire is a kind of virtual personal assistant that lives inside your telephone line and uses speech recognition to manage phone, fax and e-mail communications. To activate the programme simply say 'Wildfire' and it responds with a bright, mildly surprised 'Oh, hi!' as if it was actually glad to hear from you. Tell Wildfire to get Bob on the phone and it will dial the number automatically. Ask it to read out your e-mail messages and it reels them off in a briskly efficient yet slightly seductive female voice.

Voice operation and speech synthesis are becoming increasingly commonplace in mobile phones, automobile navigation systems and other devices in which portability and ease of use are key. Research is even under way to replace familiar desktop icons with 'earcons', audible tones that would alert users to incoming e-mails, changes in stock prices or important news bulletins. For mobile devices, this technology would allow users to keep their eyes on the road, the pavement or the person in front of them instead of casting furtive glances at a computer screen. An investment banker could hear the steady murmur of the Dow, for example: when it goes up, perhaps the high-pitched squeal of fireworks; when it goes down, who knows, perhaps the sound of a flushing toilet.

Despite the success of these services, though, speech recognition is still not yet ready for prime time. Wildfire and Tellme only work on the telephone, where the software is not distracted by other voices or random noises.

Noise is, in fact, the main obstacle to the more widespread introduction of hearing computers. So far, speech-to-text software programmes – like IBM's ViaVoice and VoiceXpress, made by the now bankrupt Belgian technology firm Lernout & Hauspie, which convert dictation into written text – only perform well in unnaturally quiet environments. Most of the software is also speaker-dependent, meaning it is only compatible with a single user's voice, and even then only after extensive training.

The training sessions themselves normally involve reciting bland texts into a microphone until the system becomes familiar with an individual's distinctive accent, pronunciation and other vocal idiosyncrasies. The system uses neural networks to learn the user's speech patterns and store them in its database. Though the technology is improving steadily – L & H, for example, claimed that its VoiceXpress continuous speech-recognition software took just five minutes to train and could achieve 98 per cent accuracy – there's still a long way to go before speech recognition will work in the real world of noisy offices and busy streets.

Jupiter, a phone-based interface developed by Victor Zue and colleagues at the MIT Laboratory for Computer Science's Spoken Language Systems group, is one of the best listeners around. Dial up Jupiter and ask for the temperature in New York and it will give it to you. How about Paris? Jupiter has that information too. Need to know the humidity in Tokyo or the weekend forecast in Rio? Jupiter can tell you. 'Speech recognition enables our machines to match our own communication abilities,' Zue says. 'But to become a truly human-like conversation partner, the machine needs to do far more than just hear and recognise words; it needs to understand them.'

Master of a vocabulary of about 1500 weather-related terms, Jupiter doesn't understand much – just enough to hold forth about the weather. To do that Jupiter parses human speech into five easy pieces. First, a speech-recognition program converts the spoken sentences into text. Then a language-understanding program breaks down the sentences grammatically to identify subjects, objects and verbs. Once the text's meaning is understood, an information-retrieval program fetches the requested information from US National Weather Service reports on the Internet. A language-generation program then composes a text sentence that presents this data in the user's preferred language, and finally a speech-synthesis program converts the text into computer-generated staccato speech. And all this happens within the space of a few seconds.

In some respects, speech recognition systems like Jupiter emulate how language is processed in the brain; but in human hearing, prosody – the varying rhythms and intonations of speech – is crucial to unlocking meaning.

The first task in recognising and interpreting spoken language is to decide where one word ends and the next one begins. With written text this is easy, since a blank space conveniently marks the boundaries between words; but in speech, word boundaries are usually blurred and often non-existent. So the brain tries to detect prosodic patterns that distinguish one word from another – things like changes in volume and tone of voice, pauses for emphasis, drawn out pronunciations. Neurons in the brain respond to these patterns so that listeners can quickly decipher what is being said. Studies have shown that infants as young as three months attend to prosody to understand their parents' babbling.

In English, for example, most words are stressed on the first syllable. So when the brain picks out a pattern of stressed and then unstressed syllables from the continuous flow of speech, this is strong evidence that those sounds are in fact a word. When the brain latches on to what it suspects is a word, the next task is to decide just which word it is, and what it means. When you hear the first syllable of the word 'weather', for example, your brain quickly rushes through a lexicon of possible matches. If you only have the 'we' sound in 'weather' to go on, the word could be 'wet', 'welcome', 'Wednesday' or 'wedding', among many others. The brain rummages through all of these possibilities, even the ones that are meaningless in the context of the entire utterance.

To choose the right word – and the right meaning – the brain then analyses larger units of speech, called 'breath groups' or 'intonational phrases'. Placed in this wider context, the individual words and their collective meaning become clear. But it's the prosody of spoken language that drives the brain's analysis.

These are some of the same kinds of things that Jupiter is listening for when it's fielding questions about the weather. So far the weather is all it can talk about, but Zue and his team have developed similar interfaces that handle flight schedules, online classified ads for automobiles and information about traffic and landmarks such as museums and universities in Boston, Massachusetts.

Voice-interaction like this is crucial if mobile Web access is ever to become practical on a widespread basis. Currently, mobile technology is hampered by the need to compress more and more functions into smaller and smaller handsets. As Zue says, 'You can't type with toothpicks.' If mobile devices such as cell phones, personal digital assistants and even laptops were equipped with speech recognition, users could place calls, send e-mail and make purchases or reservations through simple verbal commands.

Even Zue admits, though, that noise could well drown out even the most carefully articulated commands. What can be done? Well, you could shout, but that's unlikely to go down well in public places. Or you could do what people always do in loud environments such as crowded subway stations: 'Put a face on the interface,' says Zue.

TALKING HEADS

Jackie Strike, Kiwilogic's virtual contender in the 2000 US presidential election, is one of the best-known faces on the Web. An award-winning war reporter in Vietnam as well as being successful in business before entering politics, Jackie Strike campaigned from a website to become leader of the free world. Needless to say, she came a distant third to George W. Bush and Al Gore.

Gazing imperiously from behind a lectern, Strike took on the issues of the day, railed against alleged irregularities in the Florida vote and responded to written questions in a husky robotic monotone. Ask Jackie Strike about social security (the American state pension system), gun control, the environment or universal health care, and the candidate replied with brief but impassioned and surprisingly well-informed opinions.

Though lingubots such as Jackie Strike are often called intelligent agents, they are really not that smart at all. The agents don't think for themselves, but are programmed to respond to a repertoire of key-words and questions with a set of prerecorded answers. Jackie Strike knows a lot about US politics, for example, but very little about anything else. Like many actual politicians, Strike is just regurgitating what a teleprompter tells it.

When Jackie Strike fields a question on gun control, for example, it has a detailed and thoroughly credible answer, not because it's intelligent but because programmers expected the question and deposited the reply in the database. Strike's programmers are like a politician's spin doctors: they try to anticipate all the possible questions that could be raised about the candidate – What's her position on gun control? Has she ever fiddled her expense accounts? Did she or didn't she inhale? – and then come up with plausible answers.

The difficulty lies not in thinking up all the different questions a user might ask, but in getting the lingubot to discern all the various ways a single question can be posed, such as, 'What's your position on gun control?' 'Do you support gun control?' 'Should children carry guns?' or 'Where do you stand on guns?' To save time, Kiwilogic developed a knowledge database of more than 1000 predefined rules that cover the most frequently asked questions. When a question is submitted, the lingubot parses the query, identifies keywords and then looks up the appropriate answer in its database.

In this way a lingubot can become an expert on almost anything. Take Marc, another Kiwilogic lingubot, which was built to explain Eye-Trek, the cordless television system from Olympus described in the chapter on 'Sight'. Since Olympus believes that people who might be interested in Eye-Trek might also be interested in science fiction, Marc has a database with detailed information on 150 of the most important science-fiction films. Marc monitors its own performance, so if it is blindsided by a question to which it doesn't know the answer – for instance, in the 1951 sci-fi classic *The Day the Earth Stood Still* what did Patricia Neal say to the robot Gort to prevent it from destroying the planet? – it will store the query and tell the programmers to come up with the correct reply.

The next generation of lingubots will be able to do more than just recite prechewed answers; if they don't know something, they will be able to scoot out onto the Web to forage for the answer. If Jackie Strike is flummoxed by a detailed question regarding social security, it can waffle on a bit to win time while it's out scanning the Web for data that are relevant to the question.

To conduct that search, Jackie Strike might well use Kenjin, made by the British software firm Autonomy. Once installed on your PC, Kenjin, which means 'wise man' in Japanese and operates as a window at the bottom of your screen, watches and understands what you do and automatically suggests links to related websites and other files on your hard drive. Kenjin not only analyses the documents on your screen, but also extracts concepts from the text and determines which ideas are the most important. The software then identifies similar concepts and ideas in other sources. So rather than having to stop what you're doing to roam the Net in search of more information, Kenjin analyses what you need and brings the information to you.

For example, let's say Jackie Strike is asked about its relationship with Leonardo – the boyish movie star, not the inventor, painter, poet and practitioner of mirror writing. As Strike launches into a discourse on the definition of the term 'sexual relations', Kenjin is out scanning the Web for relevant links, e-mails and documents about DiCaprio, which then automatically appear ranked in order of relevance. Because Kenjin understands the context of the discussion rather than simply identifying an ambiguous keyword, it sends lots of stuff about DiCaprio and nothing at all about Da Vinci. Candidate Strike processes this information and can cite tabloid revelations about the actor's other romantic interests while assuring listeners that the relationship is strictly platonic.

Over time, Kenjin gradually learns about the subjects that interest the user and applies this knowledge to further refine and personalise the information it offers. A lingubot equipped with the charisma of

Jackie Strike's programmers and Kenjin's intelligence could be a real vote-winner.

'Lingubots don't need to be as intelligent as humans,' says Von Wendt. 'They just need to answer practical questions and remove the burden of looking for information from people. The Web will be much more fun when it can hear and talk.'

Fun is definitely one of the primary motivations behind synthetic interviews, an interactive video technology developed by Scott Stevens, Michael Christel, Alex Hauptmann and Donald Marinelli at Carnegie Mellon University's Entertainment Technology Center (ETC) in Pittsburgh. The ETC's first synthetic interview is a question-and-answer session with Albert Einstein. Using a three-dimensional 'real image' video projection system, Albert Einstein, played by actor Jerry Mayer, hovers before viewers. Users can converse with the virtual physicist through a speech-recognition and digital video-information retrieval system. The effect is something akin to a Madame Tussaud's exhibit come to life – without the wax figure but with about the same degree of uncanny realism.

To chat with Einstein, the user poses a question, such as: 'Why do you refuse to accept the validity of quantum physics?', by speaking into a microphone. The speech-recognition software analyses the query and scans its database of Einstein's actual writings, lectures and speeches until it finds the right answer. The database contains a vocabulary limited to certain specific topics, which incorporate phrases, terms and proper names – such as 'relativity', 'atom bomb', 'Leonardo da Vinci', 'Isaac Newton' and 'Nobel Prize' – that are strongly associated with Einstein. Once the correct reply is located, the appropriate video clip is called up and Jerry Mayer comes back with why he thinks God does not play dice.

'A synthetic interview is really a big *Jeopardy* game,' says Marinelli, a professor in the School of Drama at the university's College of Fine Arts. 'It's a database of answers in search of the proper questions.' *Jeopardy*, first introduced in the 1960s, is one of the longest-running and most popular American game shows on television. Instead of asking questions the host of *Jeopardy* provides answers, to which the contestants must supply the corresponding query. In reply to the answer: 'This is one of Einstein's most famous equations,' for example, a contestant might respond with: 'What is $E = mc^2$?'

The synthetic interview with Einstein involves hundreds of pre-recorded video clips that show Mayer answering questions. To make the experience as lifelike as possible, Marinelli and crew also depict Einstein in various contemplative, humorous or active poses – scenes of the actor jotting down notes, scratching his head, drinking a cup of coffee, laughing at a private joke or motioning the user to come closer – that are shown during the down time between sessions. 'The speech-recognition component allows the user to become a protagonist, to control the direction and tone of the interview,' says Marinelli.

To make these encounters even more up-close and personal Marinelli and the ETC team are creating Doc Beardsley, a physical robot with the mannerisms of an absent-minded professor. At the moment Doc Beardsley is just a mechanical head, but the Carnegie Mellon researchers hope to add an arm in the near future and, eventually, an entire body. Doc Beardsley's facial expressions, which range from reflection and puzzlement to insight and astonishment, are created by a network of servo-motors and actuators that can raise an eyebrow in surprise or pull back the lips into a smile. Users interact with the good doctor in much the same way as the virtual Einstein: ask a question and Doc Beardsley's search engine fetches the answer and choreographs the appropriate facial cues as it delivers its response.

The robot's absent-mindedness is not just a charming touch of verisimilitude; it allows Doc Beardsley to go off at amusing but irrelevant tangents when it doesn't find a matching reply in its database. These kinds of built-in idiosyncrasies are important because, unlike lingubots, Marinelli wants synthetic interviewees to provide meaningful contact between the user and the recreated personality, not just assistance in making a transaction.

Synthetic interviews are not limited to the legendary figures of science and absent-minded professors. The ETC's first commercial products will likely feature celebrities rather than great thinkers. Marinelli imagines synthetic interviewees from the pop-culture firmament starring on talk shows, call-ins, websites and CD-ROMs.

Synthetic interviews may have more domestic uses as well, perhaps as a form of interactive home video. Parents could create synthetic interviews for grandchildren or great-grandchildren they might never see. Instead of sitting down with piles of photo albums, families of the future could gather around a virtual grandma or grandpa to hear stories about what their lives were like in the twentieth century. The grandparents could cross-link their own audio and video recordings with relevant snippets from home videos and even news broadcasts, so that their descendants had a complete personal and historical overview of decisive moments in family history.

Artificial-intelligence researcher Hans Moravec imagines an even more sophisticated version of the synthetic interview, one that contains more than just a few clips of fond reminiscences. 'A kind of portable computer ... is programmed with the universals of human mentality, your genetic makeup, and whatever details of your life are conveniently available,' Moravec writes.

> It carries a program that makes it an excellent mimic. You carry this computer with you through the prime of your life; it diligently listens and watches; perhaps it monitors your brain and learns to anticipate your every move and response. Soon it can fool your friends on the phone with its convincing imitation of you. When you die, this program is installed in a mechanical body that then smoothly and seamlessly takes over your life and responsibilities.

The technology still has a long way to go before it's realistic enough for users willingly to suspend their disbelief, regardless of whether the dialogue is with Madonna or the user's deceased grandmother; and it will be many decades before Moravec's vision becomes practical. But Marinelli is convinced that the more lifelike these robots become, the more useful they will be, both as sources of entertainment and as virtual time capsules of personal history. 'Computer science needs illusion,' he says. 'That's why theatrical tricks are so important to making the user's experience genuine. People will believe that they are dealing with a thinking, sentient being.'

Michael L. Mauldin, a computer scientist at the Language Technologies Institute, also at Carnegie Mellon, is injecting bits of emotional intelligence into computers with 'verbots', or verbal software robots, that understand and speak English through a combination of speech recognition, voice synthesis, artificial intelligence and real-time animation. Made by Virtual Personalities, Inc. in Los Angeles, verbots are animated characters that inhabit your computer screen and come with a repertoire of prefabricated facial expressions. 'If the computer is puzzled, its face should show a quizzical look and it should say, "I'm confused. Can you explain?" ' says Mauldin. 'Emotions are another channel of information flow, and different facial cues will trigger different responses in the user. Instead of just looking at a screen and clicking on a mouse, we can interact with computers on an emotional level.'

Mauldin, founder of Internet search engine Lycos and creator of Julia, one of the earliest and most convincing virtual agents, is convinced that the two keys to creating a successful interface are trust and comfort, so all his verbots are relentlessly average. To ensure averageness, verbots look, well, average – they are not preternaturally attractive, like the virtual newscaster Ananova. 'We don't want to make people uncomfortable,' Mauldin says. 'If you feel like you're talking to Cindy Crawford all the time, you might behave differently than the way you actually are.'

According to Mauldin, the most important service a verbot can provide is to shift the burden of interaction from the user to the computer. If you hit a snag in word processing or encounter a glitch during an e-commerce transaction, instead of getting that annoying error message, a verbot could step in to offer advice on what the problem is and how to solve it; and the verbotic voice would be more emotionally nuanced than Jackie Strike's patter. If your laptop battery is running low, for instance, the verbot might gently remind you to recharge it when you get home. If, on the other hand, a virus is eating its way through your hard drive the verbot would take on a much more urgent tone. Mauldin argues:

> People are used to telling other people what to do. Why can't that be the way people interact with computers, too? Instead of spending months learning to master basic computer skills, we can talk to a computer and it can talk back. People are comfortable talking to a machine, as long as they know that it's a machine. As speech-recognition and speech-generation technology improve, the question 'Excuse me, but are you human?' will become common.

SPEAKING IN TONGUES

Such a question could only have been imagined by Wolfgang von Kempelen, a eighteenth-century Viennese engineer and linguist who published one of the first detailed descriptions of a speaking machine in his 1791 book *Mechanismus der menschlichen Sprache nebst Beschreibung einer sprechenden Maschine.* Although Von Kempelen's device could not hear, it could articulate complete words and short sentences by means of a hand-operated bellows and a series of levers. The actual speech-production mechanism, contained inside a wooden box, was disarmingly simple: a rubber nose and mouth connected to the bellows via an oscillating reed.

To make the box talk, airflow from the bellows was conducted into the mouth through the reed. The operator altered tone and pitch by cupping a hand across the mouth, much as a trumpeter uses a mute to muffle the sound of his horn. Consonants and vowels were produced by manipulating the levers: two levers produced fricatives, letter sounds produced by forcing the breath through a narrow opening between the teeth, while a third made a rolled 'r' sound by dropping a wire onto a vibrating reed. The pronunciation of the letters 'z' and 's' was given added realism thanks to hissing whistles. In one of the voice box's more awkward requirements, the operator had to plug the nostrils with two fingers to prevent the production of nasal tones. According to Hartmut Traunmüller, a professor at Stockholm University who had the privilege of playing Von Kempelen's still extant device in 1997, the voice is 'similar to that of a child and quite loud'.

Artificial speech became more fluent around 1835 with the Euphonia – or the Amazing Talking Machine, as it was also known – made by Joseph Faber, a German immigrant to the United States. Using the same basic principles as Von Kempelen's box, the Euphonia sported a tongue and throat whose shape could be altered to produce different sounds. The apparatus was controlled via a keyboard, while the bellows was operated by a foot pedal. This talking machine was truly amazing because it could not only speak several European languages but also sing, once treating astonished listeners in London to a rendition of 'God Save the Queen'.

Figure 5. Joseph Faber's Euphonia.

An electronic version of the Euphonia – the Voder, created by Homer W. Dudley at Bell Labs in the United States – made its debut at the 1939 World's Fair in New York. Apart from the electricity, it is remarkable how little the technology changed in the century or so between Faber's device and Dudley's Voder. As the author of a cover story in a January 1939 issue of *Science News Letter* wrote,

This new synthetic orator . . . is a compact machine resting on a small table. No recording of any kind is used in this latest addition to the anatomy of the Mechanical Man. A girl at a keyboard controlling varying electrical currents does the trick. It has a pair of keyboard units, more than a dozen other controls and an electrical circuit featuring a vaccum [*sic*] tube and a gas-filled discharge tube. Seated at the keyboard of the 'Voder', this young lady can carry on an ordinary conversation by pressing keys . . . A pedal operated by her right foot enables the operator to make voice inflection, and synthesised speech is heard from a loudspeaker.

Scientists at Bell Labs are also responsible for introducing synthetic speech to pop culture. In 1962, Bell Labs researcher John L. Kelly taught his Vocoder synthesiser to sing 'Bicycle Built for Two'. Science-fiction author Arthur C. Clarke heard the Vocoder in action during a visit to the lab and later had HAL, the psychotic computer in *2001: A Space Odyssey*, sing the same tune in his novel and screenplay.

After their appearance on the big screen, talking machines really hit the commercial mainstream in 1978 when Texas Instruments released Speak & Spell, the first device in which the human voice was electronically duplicated on a single chip. This educational toy for seven-year-olds and up had a computerised voice that pronounced words aloud; children then tried to spell the words on an alphabetically arranged keyboard. The computer informed players if they had spelled the word correctly. Speak & Spell was a novelty when it first appeared, but today toys that can speak and hear, such as computerised pets like Sony's Aibo and robotic dolls like the US hit My Real Baby, are commonplace.

Like HAL, the Babel fish is another fictional piece of talking high tech that's several giant leaps closer to reality, this time thanks to Alex Waibel and colleagues at Carnegie Mellon University's Interactive Systems Laboratory. As described in Douglas Adams' 1979 book *The Hitchhiker's Guide to the Galaxy*, the Babel fish, properly inserted in the ear, enables you to 'instantly understand anything said to you in any form of language'. Waibel has come up with a way for intergalactic travellers to wear their tongues on their sleeves rather than in their ears.

LingWear is a spoken language travel assistant consisting of a laptop PC that's light enough to carry in a knapsack, a microphone that clips onto a shirt or collar, and a display and loudspeaker unit worn on the wrist. When users find themselves in foreign parts where they don't speak the language, the LingWear provides real-time translations within certain specific subject areas, such as travel-related conversations or dialogues having to do with simple medical conditions. The system is currently fluent in English, German, French, Italian, Japanese and Korean.

Say you're checking into a hotel in Tokyo and have a slight headache from the flight. Ask the concierge in English if he has any aspirin and the LingWear translates the question into Japanese and broadcasts it from the speaker on your wrist. When the concierge replies, the

LingWear hears his response and automatically translates back into English. Should you be wearing a heads-up display like Steve Mann's as described in the chapter on 'Sight', translations can also be viewed as subtitles under the face of the person to whom you are speaking.

The LingWear gets its linguistic skills from the Janus speech-translation system. Janus is a speaker-independent system that translates conversations about appointment scheduling, hotel reservations and travel planning in much the same way as Victor Zue's Jupiter. The system has a vocabulary of about 10,000 words, can access train schedules, calendars or personal agendas to get additional information, as well as provide location information through a global positioning system (GPS) connected to the laptop. The device can even serve as a kind of aural tour guide, providing historical background commentary during museum tours or brief descriptions of architectural highlights in selected cities.

To provide real-time translations, though, Janus has to handle the messy and imprecise locutions that make up everyday chatter. Oral speech, at least as spoken by human beings, rarely obeys rigid syntactical rules. Instead, spoken words are punctuated by pauses, hesitations, repetitions and non-sequiturs – all of which are perfectly comprehensible to a human listener but completely flummox a computer. So instead of a literal, word-for-word translation, Janus interprets and summarises what it thinks a person means.

This is accomplished through an interlingua, an intermediate computer language that describes the intention of an utterance independent of the language in which it is expressed. Imagine you're still in Tokyo and trying to make an appointment with a Japanese colleague. If she suggests you meet on Monday and you say, 'Monday is awful,' Janus understands that this means Monday is an inconvenient day for the appointment, not that you have an aversion to that particular day of the week. Rather than render this word-for-word in Japanese, Janus simply translates what you mean: 'I can't make it on Monday.'

Machine translation programs like Janus have applications apart from just travel and tourism. LingWear devices could enable people in the developing world, who may not have access to proper health care, to consult in their own languages with physicians who may be thousands of miles away. The translators would also come in handy during peacekeeping operations in volatile regions like the Balkans, where good communication between soldiers and civilians is essential to maintaining stability. Handheld machine translation devices like LingWear could even help slow, if not reverse, the process of language extinction.

Of the roughly 6500 languages now spoken, up to half are already endangered or on the brink of extinction. Linguists estimate that a language dies somewhere in the world every two weeks. Languages, like all living things, depend on their environment to survive. When they die out, it is for reasons analogous to those that cause the extinction of plant and animal species: they are consumed by predator tongues, deprived of their natural habitats or displaced by more successful competitors. In this type of linguistic natural selection, though, the survival of the fittest is not determined by intrinsic merits and adaptability alone; the economic might, military muscle and cultural prestige of the country in which a language is spoken play a decisive role. For languages without the clout of English – in which three-quarters of the world's mail and up to 80 per cent of e-mail is written – technology may lend a helping hand.

Waibel and his colleagues believe portable devices like the LingWear could help speakers of threatened languages interact with speakers of dominant tongues without the threat of losing their linguistic and cultural identity. 'Non-English Web pages are surpassing English Web pages in number,' says Waibel, 'which shows that the Web actually has the potential to promote diversity rather than the linguistic domination of English, as is often feared.'

Speaking machines have come a long way since Von Kempelen's wheezing voice box, and one of the best places to hear what they can do was, until the company went bust, at the headquarters of Lernout & Hauspie, formerly located in Ypres about an hour south of Brussels. Belgium is a country riven by linguistic division – the northern, Flemish-speaking part of the country is often barely on speaking terms with the southern, French-speaking part – so there was a satisfying irony to the fact that a Belgian firm manufactured some of the most natural-sounding speech-generation systems around.

Lernout & Hauspie's RealSpeak text-to-speech software, which reads aloud text in a strikingly human voice, sounds so realistic because it's composed of snippets of actual human speech, rather than synthesised computer speech such as the less than dulcet tones of

Jackie Strike. To achieve this level of verisimilitude, though, requires a lot of talking, and it doesn't come cheap.

The RealSpeak engine stores samples of human speech – from large phonetic units such as words or syllables (such as the 'cen' in 'central') down to the smallest units called diphones (sound units containing a combination of two sounds such as the 'al' in 'central') – in a database. To compile this database, hours of carefully chosen words, sentences and paragraphs are first recorded by a native speaker. Subsequently, a large amount of linguistic information is added to the recorded speech: boundaries between sound units are marked and stressed syllables are identified. This is currently a semi-automated process that requires an extremely patient person to go through the entire data set to verify whether the annotation that has been automatically generated is correct. Even for a limited database of 5000 words or so, this is a daunting and time-consuming task. It took one L & H employee several months to complete this task for a single German-language database.

Once a complete phonetic library has been compiled, the Real-Speak engine applies language-specific rules and prosody models to produce natural-sounding intonations for complete sentences and phrases. When the computer reads a text, the RealSpeak engine recognises the sound units in the appropriate context, retrieves the corresponding phonemes from its databank, arranges them in the proper order and converts the text into speech.

Artists, being the early adopters that they are, could use this kind of technology to create an entirely new art form – or at least a compelling new form of personal entertainment. German composer Richard Wagner wanted to create the *Gesamtkunstwerk*, or 'total artwork', a production that combined theatre, music, dance and verse to involve all art forms in a single performance. Wagner attempted to transcend the artificial boundaries he felt separated works into different categories: music was only for the ears, painting was only for the eyes, poetry was only for the mind. In his operas he tried to meld these forms into a single piece that would engage all the senses in spectacle.

A new kind of *Gesamtkunstwerk* might now be possible, thanks to speech recognition systems and the other sensory technologies described in this book. Just as hypertext brings an extra dimension to the written word by connecting it to other texts on the Web, 'hypersense' could bring an extra dimension to art and entertainment by connecting all the senses in a single interactive experience.

We've already seen in the chapter on 'Sight' how devices like the VRD make virtual environments more immersive and compelling. In this chapter we have heard how speech-recognition systems enable users to interact with virtual characters. Imagine what would happen if smell, taste and touch were added to the sensory mix. As we'll discover in the rest of this book, it's now possible to transmit fragrances, flavours and tactile sensations through electronic devices. So you could sit down to a synthetic interview with your deceased grandmother and not only see her face and hear her voice, but also smell her perfume, taste her cooking and feel the texture of her skin. The technologies that make these sensations possible are discussed in the chapters that follow. Computer-game designers, film-makers, interactive artists and playwrights could use these devices to dream up scenarios that titillate all the senses and in which the user him- or herself is the centre of the action, just as Woody Allen inserted himself into crucial moments of history in the film *Zelig*.

When I was a boy I devoured the 'We Were There' novels, a series of adventure stories about important eras in American history – the Revolutionary War, the exploration of the West, World War II. Each story revolved around the experiences of a boy and a girl who somehow got caught up in these great moments that shaped the US. The titles emphasised this personal dimension: *We Were There on the Oregon Trail*, *We Were There at the Battle of Gettysburg*, *We Were There in the Klondike Gold Rush*. Hypersense dramas will be like 'We Were There' stories – but for real.

Facial mapping technologies already allow individuals to perform in music videos, films or online games. Using just two simple digital mug shots – one frontal portrait and one side view – Digimask, a UK firm based in Richmond, Surrey, can create a virtual three-dimensional clone of any face in minutes. The images are scanned into a computer, where a facial-cloning program maps the facial features onto a virtual cranium. After a few nips and tucks with the cursor, the new (though not necessarily improved) you can perform all sorts of amazing tricks, including executing 360-degree turns, spinning around upside-down and displaying lifelike smiles, grimaces, snarls and leers.

While seeing your own stupidly grinning face spinning through cyberspace might be amusing, Digimask thinks its virtual visages have some novel personal-entertainment applications. Teenagers can attach their heads to virtual bodies and insert themselves into music videos to snog their favourite pop stars. Gamesters can put themselves where the action is by attaching their own heads to nubile young superheroes, daring racing car drivers or obnoxious professional wrestlers. Add the full complement of sensory technologies, and users will be able to see, hear, smell, taste and touch everything that happens to them. Imagine it: you could be there – in the latest Spielberg film, in Madonna's video, in your favourite sitcom – and have all the sensual experiences to go with it.

If the era of the digital *Gesamtkunstwerk* ever does arrive, it will enable every user to become the screenwriter, director and star of his or her own hypersense drama. Fame will no longer be cruelly limited to a fleeting fifteen minutes on some daytime talkshow or *Big Brother*-like competition. Fun, celebrity, romance and adventure could be only as far away as the nearest hypersense virtual-reality environment. To explore some of the other technologies that could make hypersense a reality, we now turn to the most evanescent of senses: Smell.

SMELL
Adventures in Odour Space

'The invention of odours ... was aimed at making us rejoice, exciting us and purifying us so as to render us more capable of contemplation.' MICHEL DE MONTAIGNE

The next time you're in London go to Jo Malone, the classy purveyor of fragrances, bath oils and body lotions near Sloane Square, and enter one of the shop's fragrance-testing pods. The pod is about the size and shape of a telephone box and looks like one of those soundproof booths used in 1950s television quiz shows. Once behind the all-glass door, you are isolated from the outside world in a little micro-environment of your very own.

Inside the pod is a touch-sensitive video screen displaying a menu of Jo Malone scents – Honeysuckle and Jasmine, Verbenas of Provence, Tuberose, Lime Basil and Mandarin, among others. The only sound is the low but insistent hum of the ventilation system, which streams a jet of air into your face that is both gentler and more diffuse than the blast you get from those adjustable nozzles above airplane seats.

Select a fragrance – Lime Basil and Mandarin, a bestseller – by touching the screen and that scent quickly fills the booth, as a grainy black-and-white video with a bass-laden soundtrack displays a beautiful young couple frolicking in a bedroom while brandishing a variety of Jo Malone products. Then, as quickly as it came, the smell of Lime Basil and Mandarin dissipates.

This intimate olfactory moment is brought to you by Aerome, a Düsseldorf-based company that, according to the information on its website, 'offers advertisers push-button access to the world of scent'. Jo Malone's fragrance-testing pod is equipped with an Aerome

MediaScenter, a touch-screen computer that is coupled to the firm's ScentController technology.

The ScentController is, in effect, a video player for the nose. Insert a ScentCartridge – a white rectangular box that looks very much like an ordinary video cassette but contains six prefabricated scents – into the device, and it releases the chosen aromas in sync with scenes from the Jo Malone advertisement. The amount, intensity and duration of the smells can be precisely controlled, and Aerome claims it can create any desired fragrance. New technologies like the MediaScenter are able not only to create odours and disperse them into the air, but transmit them over the Internet as well. 'We are reaching beyond hearing and sight to a third dimension in the world of communication,' enthuses Marc Meiré, Aerome's founder.

Right now Meiré's reach may well exceed his grasp. The ScentController is an impressive dispenser of odours, but it's not necessarily better than simply stepping outside the fragrance-testing pod, twisting off the top of a bottle of body lotion and inhaling deeply. Yet Meiré, along with a host of other advertisers and retailers, has got wind of the fact that smell sells, and Aerome is just one of a clutch of new firms using the power of the PC to prove that the quickest way to a shopper's cash is up her nose.

Smells can subtly influence mood and behaviour, so many firms are introducing made-to-order odours into shopping environments – the scent of freshly baked bread in supermarkets or the smell of rich leather in car dealerships – to entice consumers to linger, thereby propping open the window of opportunity for a sale. The air in the BA lounge at Heathrow Airport is regularly refreshed with the tangy scent of the sea and the smell of freshly cut grass, and BOC Gases in Guildford, Surrey, has experimented with the bracing aroma of newly washed linen in Thomas Pink, the famous shirtmakers in London's Jermyn Street. Interest in designer smells is so strong, in fact, that some companies are even considering the creation of corporate odours to go along with their corporate logos.

PCs like the MediaScenter can do more than just pass gas: they can smell, too. Electronic noses, arrays of odour-sensitive electrochemical sensors linked to high-powered computers, have been on the market for the past several years, used primarily to trace explosive residues, analyse blood-alcohol levels and carry out quality-control tests in the food and beverage industries. A new generation of e-noses is beginning to replicate the speed, sensitivity and discrimination of the human nose, abilities that will enable a digital proboscis to do everything from assist in medical diagnoses to identify leaks of hazardous substances.

Thanks to these new engines of olfaction, your family physician could well be making house calls again — that is, giving a preliminary diagnosis based on information gleaned from an electronic nose embedded in your phone — and fragrant websites, scented e-mails, odoriferous interactive games and aromatic online advertising may be coming soon to a computer screen near you.

Aerome's ScentController system provides a whiff of things to come. Based on the work of Hamburg inventor G. Ulrich Wittek, the technology was originally developed as part of an effort to add olfaction to the movies, an endeavour that makes Wittek part of a venerable technological tradition stretching all the way back to the ancient Romans. The stages of Roman amphitheatres were often sprinkled with aromas or scented with thin mists of perfumed water during performances.

In the 1950s, Swiss professor Hans Laube invented Smell-O-Vision, a machine installed in movie theatres that emitted puffs of specific odours in synchronisation with the action on the screen. Like a soundtrack, Laube's scenttrack — introduced in the 1960 film *Scent of Mystery*, starring a paunchy Peter Lorre and the young Denholm Elliott, the British character actor who later found fame and fortune as Harrison Ford's sidekick in the Indiana Jones movies — was controlled by a 'smell brain' that contained an assortment of odours such as coffee, garlic, cigarette smoke and shoe polish. These aromas were pumped into the theatre through a network of hidden plastic tubes attached underneath each seat.

According to a reviewer from *Time*, the smell brain worked like a dream: 'The customer ... gets a snootful: apples, peaches, brandy, wine, tobacco, shoe polish, peppermint, roses, garlic.' Getting the smell of garlic and shoe polish into theatres turned out to be relatively easy; getting them out again proved decidedly more difficult. The Smell-O-Vision's concoctions lingered far too long, creating a foul-smelling olfactory fog. As a result, audiences and critics alike turned up their noses at the device. 'Most customers will probably agree that the smell they liked best was the one they got during intermission: fresh air,' the *Time* reviewer wrote.

Despite the bad reviews, it seems the smellies are back again. Aerome has solved the malingering problem by ensuring that only tiny amounts of scent are released in any one burst. When a Jo Malone patron touches the screen in the fragrance-testing pod a signal is transmitted to the computer, which responds by directing a jet of air across the top of the ScentCartridge, where the fragrances are stored in special odour-absorbent granules. The ScentController plays the cartridge like a clarinet, opening and closing an array of valves to release the correct scents in time to the video and music. The user feels a slight breeze on her face and enjoys the selected fragrance for just a few seconds before it fades without a trace.

Aerome is looking to place its MediaScenters in well-trafficked retail locations and is working with cosmetics companies, fast-food outlets and automobile manufacturers to introduce larger-scale e-commerce and entertainment applications. The company has already spun off an advertising agency that sells scented online banner ads. One of the new firm's first clients is the German drugstore chain Kaiser's, which has placed 500 Internet kiosks equipped with Aerome ScentControllers in its stores across Germany. Customers can surf the Web, search for product information and click on banner ads at the kiosk. After clicking on the banner, a scented commercial appears that's redolent of the advertiser's chosen fragrance.

Aerome was also in the air at the EXPO 2000 in Hanover. The firm brought the 'official Expo welcome scent' and the smells of the five continents to life through the terminals of the visitor information system and offered an olfactory film festival of forty-five movies about the Expo themes of man, nature and technology. The film about forest fires, for instance, was accompanied by the smell of smoke.

IN THE REALM OF THE SCENTED

In 'Song of Myself' Walt Whitman praises 'the scent of these armpits aroma finer than prayer', easily one of the most sublime and ridiculous lines in all of American poetry. Whitman's glorification of his personal aroma may seem odd today, given that a considerable industry has sprung up to eliminate body odours from public places. (Apart from products that actually reduce perspiration, the term 'deodorant' is a misnomer, since these concoctions don't so much de-odorise as re-odorise by masking one scent with another.) But the Whitman quote does beautifully express the intensely personal and paradoxical nature of smell.

Smell is both the most ethereal and the most visceral sense, the most primitive and the most mystical. Odours are elusive, transcendent. They evaporate. But smell something putrid and the response is intensely physical: your nose wrinkles in disgust and you recoil involuntarily as if you had put your hand in a flame. Smells are fleeting, continually crossing borders – from outside to inside, from personal to public, from material to immaterial – one reason why incense is a perfect religious symbol for the transition from the profane to the sacred.

Despite their evanescence, smells form the basis of some of our most durable and potent memories. Marcel Proust remarked on these contradictory qualities of smell (and taste, for that matter, which we will come to in the next chapter) in *Remembrance of Things Past*:

> When from a long-distant past nothing subsists, after the people are dead, after the things are broken and scattered, taste and smell alone, more fragile but more enduring, more unsubstantial, more persistent, more faithful, remain poised a long time, like souls, remembering, waiting, hoping, amid the ruins of all the rest; and bear unflinchingly, in the tiny and almost impalpable drop of their essence, the vast structure of recollection.

Smells can evoke such powerful recollections thanks to the brain structure that deciphers odours, the rhinencephalon, one of the oldest parts of the brain as well as the first sensory region to develop in the foetus. Babies can remember and recognise smells from the amniotic fluid in the womb, and can even discriminate by smell specific foods consumed during the last months of their mother's pregnancy.

When Whitman got a whiff of his armpits his rhinencephalon bypassed the neocortex where the functions of reason and logic are based and sent a jolt directly to the olfactory bulb, which is connected to the limbic region of the brain, the site where emotions and memories are processed. The limbic system, in turn, has a hotline to the hypothalamus, which controls the hormonal and autonomous nervous systems. This connection between odours and emotions explains why smell can induce such Proustian reveries. Because many smell messages bypass the neocortex and go straight to the brain's emotional centre, odour perception is often subliminal – the nose picks up messages that we are not aware of, which can prompt behaviours or moods such as instinctive attractions or aversions that we can't rationally or consciously explain.

It might also help explain why when describing smells the best we can manage is weak similes. 'If you ask someone to put a name to a smell,' writes Piet Vroon, a psychologist at the University of Utrecht in the Netherlands, in *Smell: The Secret Seducer*, 'he or she is often dumbstruck ... If the eye were to function like the nose, when you saw a sheet of bright red paper you would say something like "Looks furious" rather than "Sheet of red paper".' Smells leave us speechless because the olfactory epithelium, the yellowish sheet of nerve cells that detects odours, has few direct connections to the neocortex, where the language centres of the brain reside.

Because of the deep neurological links between olfaction and emotion, smell is one of our most intimate senses. We are rarely offended by the sight of someone else's face or the sound of someone else's voice, but it's easy to be affronted by someone else's body odour – even if it emanates from Walt Whitman's aromatic armpits. Among the senses, smell (and taste, too) results in the incorporation through the nose (or mouth) of foreign objects into our bodies. 'Vision reveals only surfaces and makes spectators of us all,' says David Howes, an anthropologist at Concordia University in Montreal and editor of *The Varieties of Sensory Experience*, a study of the role of the senses in different cultures. 'But smell comes from inside things and actually enters the smeller. If you smell something it becomes part of you and alters you.'

A smell becomes part of you when one of the 30,000 or so odour molecules in the world alights on a receptor in the olfactory epithelium ensconced high inside each nostril. To perceive the odour of just one of these molecules probably requires that about 10,000,000,000,000 (10 zillion) of them actually enter the nose.

The epithelium bristles with between 10 million and 100 million neurons. Each neuron sports a tuft of twenty to forty cilia, hair-like receptors that wave about in the thin mucusy sea of the epithelium like the tentacles of a sea anemone, waiting to snatch some unsuspecting odour as it passes by. The tips of the cilia are coated with lipids and proteins, some of which are thought to bind to odour molecules. When an odour molecule lands on a receptor, the neuron translates the smell's chemical signature into an electrical signal and dispatches it to the brain, which processes the information and identifies the smell.

No one is sure how many of these specialised proteins stud the neurons of the olfactory epithelium – a dog's nose may contain as many as 150 million, which is why they are such keen sniffers – but it is thanks to these that the average person is able to detect and distinguish between about 10,000 different smells.

In 1991 Linda Buck, then a postdoctoral fellow at Columbia University, identified the first odorant receptors, the proteins found on the surface of olfactory neurons that capture and identify odours, in mice. Buck's research has shown that each olfactory neuron expresses just a single receptor gene, and that neurons expressing different receptors are more or less randomly scattered around the olfactory epithelium. There are probably somewhere between 400 and 1000 different receptors in the human olfactory epithelium.

Buck and her team developed a genetic method for tracing the neural circuits of smell deep into the brain, and are investigating how the brain encodes the identities of different odours. Preliminary results suggest that different smells are encoded by different combinations of receptors, and that each receptor is part of the code for many different odorants. This kind of combinatorial coding, known as 'distributed specificity', could explain how as few as 400 receptors are able to encode and identify as many as 10,000 different smells.

According to this theory, an individual receptor is not specifically sensitive to a single odour molecule but can bind with and identify a range of different odours. Receptors all have a similar general structure except for a small unique region that fits like a key into corresponding areas on a selection of odour molecules. As odours waft across the epithelium, the intensity of the binding varies because certain receptors latch on to certain scents better than others – there is a better fit between their corresponding regions – and fire off stronger signals to the brain. The resulting binding pattern, or smellprint, is decoded by the brain as a specific odour. In biological terms, a smell equals the pattern of receptors it activates in the olfactory epithelium. So a smell is like a complex musical chord rather than an individual note.

Just how these patterns are formed is described by J. Y. Lettvin and Robert Gesteland in florid, steamy prose seldom encountered in scientific literature: 'There, on the surface of the cilium or cell wall, is a molecular trap, an ophidian proteinaceous affair that, on receiving a molecule of the right shape, coils about it in allosteric embrace. In so changing, the trap opens an ionic gate by some mechanical or electrical or chemical action, and through this gate particular ions speed, carrying electrically the glad tidings that a molecule is captured.'

Gesteland, a professor at the University of Cincinnati in Ohio, has spent the past forty years studying vertebrate olfactory systems, most thoroughly that of the northern grass frog, *Rana pipiens*. His research is focused on understanding the mechanisms by which odorants are detected in the nose and how this process enables one odour to be distinguished from another.

This kind of work is normally carried out by inserting tiny electrodes into the olfactory epithelium and measuring the electrical activity, or conductance properties, of the receptor cells. This is an invasive, inefficient and uncomfortable method at best. So instead of sticking wires up frogs' noses, Gesteland plucks a single receptor from the olfactory epithelium, stains it with fluorescent dyes that become brighter as voltage in the cell increases, lets it loose on a range of odorant molecules, and then measures the electrical current flowing through the receptor. The higher the voltage, the greater the conductance of that particular receptor; and the greater the conductance, the more that receptor is responding to a specific smell. Finally, Gesteland photographs the cells with a laser-scanning confocal microscope to produce a snapshot of the responses to odours of hundreds of separate receptors.

Even though the northern grass frog's olfactory gifts are modest compared to those of humans, coming to grips with such a complex system – in which there are tens of thousands of possible combinations between odour molecules and receptors – is still extremely difficult. In the frog, Gesteland discovered that 'almost every odour seems to affect almost every receptor one way or another ... It is as if every axon [the nerve fibre that hosts the receptors] expresses a point of view with respect to all compounds and combinations of compounds, and each axon has a separate point of view.' He found the same principle at work in rats. 'Nothing we've found in the frog is different from rat or mouse olfaction,' he says. 'Nature seems to have invented only one kind of nose for vertebrates.' Gesteland hopes that the measurement and analysis of the frog nose will provide new insights into the finer and much more sophisticated human instrument.

That instrument is currently being probed by Tim Jacob and colleagues at Cardiff University's School of Biosciences in Wales. Jacob is trying to detect and measure anosmia, or loss of smell, the old-fashioned way – by sticking electrodes up people's noses. Whether with frogs or humans, this is a crude technique – there's a lot of sneezing in Jacob's lab – so the Cardiff team is developing a non-invasive method for recording olfaction by placing electrodes on the surface of the nose and scalp. Recording from surface sites around the olfactory bulb is frustrating, though, because it's difficult to dis-entangle signals coming from the bulb itself and those emanating from other brain areas.

Jacob and others are also trying to use 'odour memory' – the easily conditioned association of a specific scent with a specific emotion or event – for therapeutic and medical purposes. Once such a link between a smell and a memory is made, it is extremely difficult to sever. I, for example, will always associate the smell of rubber bands with the Halloween masks I wore as a child. After a few hours of trick-or-treating the inside of the mask took on a very distinctive odour, a mix of plastic, condensed breath and rubber bands. Tim Betts of Birmingham University has put odour memory to work as a treatment for epilepsy. After conditioning a group of epileptics to associate the smell of aromatherapy oils with relaxation, almost all the patients in one study were able to reduce the frequency of their seizures when exposed to the smell.

A ROSETTA STONE FOR THE NOSE

The precise mechanisms of olfaction – how an odour molecule interacts with receptors to create an electrical signal that is interpreted by the brain as a smell – are still not fully understood. Though we may not know precisely how the nose works, the nose certainly knows a lot about us. This was made clear to me one day when my eldest son Gilles was about two. I noticed that he had extremely bad breath. His breath wasn't just bad, though, it was absolutely fetid, a sickly-sweet, slightly metallic odour that I had experienced on only one other occasion: it smelled very much like the rotting stub of his umbilical cord just before it dropped off a few days after his birth. The stench was all the more striking because up until this point, Gilles' breath had always been very pleasant, regardless of what he had put in his mouth.

Then, a couple of days later I noticed spots on Gilles' neck and along the small of his back and knew that he had chickenpox. It was the arrival of this sickness that I had detected on his breath. My memory of this episode is all the more vivid because, having been spared the malady in childhood, I soon developed chickenpox myself. Whereas Gilles was hardly ill at all, I was completely wiped out and stayed in bed for two weeks, listening to the blisters on my scalp crackle and pop like Rice Crispies in milk. I shudder to think what my breath must have smelled like.

The experience reminded me of just how knowledgeable a nose can be. Physicians in the eighteenth and nineteenth centuries were skilled at diagnosing patients by the smell of their breath. Today it's well known that certain ailments, such as diabetes and liver disease, produce very distinct odours.

Researchers at the Highland Psychiatric Research Foundation, Craig Dunain Hospital in Inverness, Scotland, have found that even schizophrenia can be diagnosed by sniffing a patient's breath. Because they break down fatty acids faster than other people, schizophrenics regularly have higher than normal levels of the alkanes butane and ethane in their exhalations, and these can be detected with a breath test.

George Dodd, research and development director at London-based Kiotech International, a biotechnology firm specialising in smells, wants to reintroduce the nineteenth-century practice of breath analysis into modern medicine; but instead of a doctor's trained proboscis, an electronic nose on a computer chip will be making the diagnosis.

Dodd, a native of Dublin born within sniffing distance of the Guinness brewery, is trying to replicate the nose's diagnostic capability on computer chips and then 'train' the chips to recognise specific smells. These electronic noses mimic human olfaction by means of an array of chemical sensors coupled to a computer, which carries out the analysis and identification process.

Dodd envisages small, cheap, hand-held sniffer chips that could be inserted into the speaker of a telephone so that patients with, say, cirrhosis of the liver could call up a computer once a month and hold a conversation with it while it analysed their breath. Within five or ten years there could be an electronic nose in every doctor's office and people might routinely carry around credit-card-sized devices to monitor health problems such as ulcers and diabetes. Such electronic noses, however, would need to be a great deal more sensitive than current devices. Ideally, they would be able to alert someone to trouble before they developed any symptoms. Researchers at Kiotech developed one such diagnostic device for asthmatics.

There is currently no easy way to monitor for the early warning signs of an asthma attack. So many asthmatics rely on regular doses of steroids to control the disease, a regimen that is potentially harmful in the long term; but if there was some way to provide advance warning that an attack was imminent, fewer drugs would be needed and those that were used could be administered more accurately. 'An asthma attack is like a forest fire in the lungs,' Dodd explains. 'If you can detect the first tree burning, you can throw a bucket of water on it. It is more difficult to control if the entire forest is ablaze.'

Building on research carried out at the University College Hospital, London, Kiotech came up with a tool that provides a weather forecast for the lungs. Biochemical studies of asthma have shown that the inflammation preceding an attack builds up over a period of weeks and is accompanied by high levels of exhaled nitric oxide. The Kiotech sniffer chip was trained to recognise nitric oxide's molecular structure by repeatedly being exposed to it. After sufficient exposure to the odour, the e-nose has no trouble accurately identifying it in the future.

To keep tabs on his asthma a patient could carry this device around in his pocket, periodically blowing into it. The sniffer chip would analyse the patient's breath and alert him to suspiciously high levels of nitric oxide, which could be the first brush fires indicating a bigger blaze was on the way.

If e-noses are to work, though, detailed molecular maps are needed of all the odours associated with disease. Since this will entail charting the interactions of the thousands of smell molecules in the air with the hundreds of smell receptors in the human nose, immense computing power – and immense patience – will be required.

Dodd cautions that the ability to decipher the complex vocabulary of smell – to produce what would be, in effect, a Rosetta Stone for the nose – is still a long way off. 'We do not know the odour code,' he admits. Accumulating the necessary data will be 'a long job, but possible', he says; and once the key odour molecules for specific diseases are identified, he predicts, 'a new generation of electronic noses can tune into those particular molecules' to provide powerful diagnostic tools.

One such tool is currently under development by John Kauer and Joel White, neurophysiologists at Tufts University School of Medicine in Boston. Instead of chemical detectors, Kauer uses optical sensors to mimic the way the human olfactory system identifies odours. Kauer and White's artificial nose is a bouquet of thirty-two sensing sites. Each site is coated with a polymer that's tagged with a fluorescent dye. (Polymers are natural or synthetic substances composed of large molecules made up of simpler chemical units.) When exposed to odour molecules, the polymers respond by swelling, shrinking or doing absolutely nothing, depending on the polymer's sensitivity to that particular molecular configuration. For example, in response to some odours, a large group of polymers might expand; in response to other odours, only a handful might expand. These changes, in turn, change the fluorescence at each detection site: the thirty-two clusters flicker and change colour like the distant lights of a city at night. This pattern of changes is the optical signature of the smell.

To record and observe these changes, the detection sites are continually bathed in a stream of light. As odour molecules drift across the detection sites, photodetectors record the fluorescence as it shimmers across the array. These images are then fed into a neural network, which analyses the response pattern to identify individual vapours.

Kauer and White are adapting the artificial nose for the detection of landmines. With financial support from the US Defense Advanced Research Projects Agency, they are working to make the artificial nose behave more like a bloodhound to enable it to root out explosives from the complex odour environments in which they occur. At present, a dog is still the best device for detecting landmines. But Kauer hopes to refine the artificial nose and shrink it so that it is small enough to fit inside a backpack but powerful enough to pick up explosive vapours in very low concentrations.

Cyranose Sciences, based in Pasadena, California, already makes such a portable device: the hand-held Cyranose 320. The 320, which looks like an ordinary mobile phone but has a stainless steel probe where you would normally find the antenna, employs a polymer sensor technology developed at the California Institute of Technology.

The module itself is an array of thirty-two different sensors, each of which consists of twin electrical contacts connected by a polymer film. When the sensor is exposed to a compound – just point the device's snout at the sample and press 'Run' – the film absorbs the vapour and swells up like a sponge. This increase in volume breaks the conductive pathways between the two electrical leads, thereby increasing the electrical resistance of the film. The change in electrical resistance creates a unique pattern, just as Kauer's artificial nose creates a unique colour spectrum. This pattern of electrical resistance can be compared with an odour library of other patterns in order to identify the smell in question. When the 320 isn't following a scent, the polymer returns to its normal shape, thus re-establishing the conductive pathways between the electrical contacts.

The 320 works in a way analogous to the way packs of baying bloodhounds work in those classic scenes from great-escape films: the sheriff gives the dogs a scrap of the convict's clothing to sniff before setting off after him through the marshes. Like the bloodhounds, the 320 'remembers' odours to which it is exposed by adding them to its library. When exposed to a new odour, such as carbon monoxide, for example, it can determine whether it has smelled it before. If so, it will identify the odour.

While Kauer and White's e-nose works with optical sensors and the Cyranose 320 with chemical sensors, the zNose™ uses sound waves to establish the make-up of a vapour. The device is based on an ultra-fast version of gas chromatography, in which an odorant and a stream of helium gas are sent through a heated column. The molecules in the mixture travel at different speeds depending upon their weight, so that molecules of similar weights eventually become grouped together. As the different molecular clusters pass through the column, they strike a quartz crystal. The lightest ones strike the crystal first and the heaviest ones strike it last, so the time of arrival of each cluster at the crystal is unique and can be used to identify the type of compound.

The zNose is different from other gas chromatographs because it uses a solid state acoustical detector with universal selectivity. The quartz crystal ordinarily vibrates at about 500 megahertz. But when a molecular cluster strikes the crystal, it shifts the crystal's frequency by an amount proportional to the total mass of the cluster. This shift enables the device to determine the odour's concentration.

Once the odour's intensity and composition are established, the zNose produces a 'vapour print' of that specific mixture. Vapour prints look like splotches of spilled soup – indeed, the device has been used to map the fragrances of twelve different varieties of Campbell's soup. The differences between perfumes, for example, appear in the shape of the US state of Oklahoma. To date, the zNose has been used to detect contaminants in wine and water and to determine the number of rounds fired in a gun based on its smoke discharge.

Though devices like the Cyranose 320, Kauer's optical bloodhound and the acoustic zNose are impressive, Gesteland points out that 'We're nowhere near having a good electronic nose.' He stresses that there's still a long way to go before the artificial versions can compete with the versatility of the genuine article. 'Technology has produced good chemical detectors,' he says, 'but you can't call them noses.'

A NEW WORLD ODOUR

The Cyranose 320 would certainly come in handy during a game of kodo, a traditional Japanese pastime invented in the sixteenth century that is a kind of aromatic version of *Trivial Pursuit*. The object of kodo, which is less than an art form but more than just a game, is to smell a selection of mystery scents, usually inspired by poems from classical Japanese literature, and try to identify as many aromas as possible. Eight people normally take part, plus a master of ceremonies who prepares and lights the aromatic incense and a scribe who keeps track of everyone's guesses. The entire ritual takes place in silence.

The master of ceremonies prepares the incense – normally a collection of leaves or bark pulverised and mixed with ash and honey to create a sticky, chewing-gum-like substance – mounts it on a burner and hands it to the participants. Each participant takes the incense burner with the right hand, places it in the left palm and 'listens' to the fragrance. They then write down the name of the plant or tree from which they think the scent is derived and the poem to which it alludes. After everyone has savoured all the scents, the master of ceremonies grades the answers and reveals the title of the source poem.

Researchers at Israel's Weizmann Institute might just succeed in introducing kodo to the West, but instead of passing around a monstrance filled with burning incense, players will gather around a computer screen and log on to the Internet. This olfactory entertainment will not be based on classical Japanese literature, but on ordinary smells found around the house. Think of it as an electronic scratch-and-sniff book to teach young children new words, or as a 'Name that Smell' game for the whole family. Or picture, if you will, a television cooking programme that transmits the actual aromas of the food into your living room. Or imagine travelling to an exotic South American rainforest and encountering there some strange, magnificent blossom with an absolutely ravishing fragrance. Why not take an odourgraph and e-mail it to a friend? All this and more will be possible, or so the scientists say, thanks to the 'digital olfactory camera' under construction at the Weizmann Institute.

Weizmann's technology combines the sniffing abilities of the Cyranose 320 with the odour-generating capabilities of the Aerome MediaScenter. To transmit the aromas of a cooking programme to an Internet-enabled television set, an electronic nose called the 'sniffer' would take an olfactory snapshot of the ingredients for, say, a pepperoni pizza – garlic, onions, basil, tomato sauce, or whatever. The sniffer would analyse the molecular make-up of all these scents, convert them into mathematical formulae and zap them over the Internet to the TV.

The TV would be connected to a computerised 'whiffer' for smell output, equipped with a set of cartridges in which between fifty and 200 different basic odours are stored and mixed. The whiffer could be located in the television set itself or perhaps in the remote control.

Once the TV received the digital recipe for the smells, the whiffer would translate the formulae, mix the appropriate odours and emit them in real time. The viewer would smell the pizza from the comfort of his own sofa, giving the term 'couch potato' an entirely new olfactory meaning.

Since it's impossible to take the dozens of different smells in a cooking programme – let alone the thousands in the world at large – and stuff them all into one box, the researchers behind the technology, David Harel and Doron Lancet, set out to do for the nose what primary colours do for the eyes. A colour printer, for example, can create any hue imaginable from the primary colours of cyan, magenta and yellow. 'There is no "purple" in the printer,' Lancet explains, 'but by getting the mix of the three colours right the machine is still able to make it.' Similarly, there may not be "pizza" in the whiffer but it can be created with the right blend from a sufficiently diverse aromatic palette.

Trouble is, there are no clearly defined primary odours in the world as there are primary colours. A single smell can have hundreds or even thousands of different components of varying nuances and concentrations. Phenyl ethyl methyl ethyl alcohol, for example, smells like roses, but an actual rose contains anywhere from twenty to thirty different 'notes', as professional perfumiers refer to a smell's constituent parts. 'Red is easy,' says Lancet. 'Rose is difficult.' Using a variation on the distributed specificity theory, Lancet and Harel believe they've come up with a way to 'bridge the gap between the electronic nose and the natural nose' and deliver not just a single rose but an entire bouquet over the Internet.

To do this Harel and Lancet devised a series of algorithms, mathematical procedures that produce the solution to a problem in a finite number of steps, that translate what the sniffer detects into scents close enough to the original for a human nose not to be able to tell the difference. Why attempt a mathematical answer to a biological question? Because the two researchers reasoned that it would be easier to produce an algorithmic recipe that approximated a smell's effect on the olfactory epithelium than to replicate the molecular structures of thousands of different smells themselves.

Though the precise number of receptors on the human olfactory epithelium is still unknown, let's assume there are 400. That means that each time we perceive an odour, a pattern of 400 different values is created. This pattern, or profile, represents a snapshot of the receptors' collective firing pattern, the smellprint for that particular odour. As far as the brain is concerned, the profile *is* the odour.

In his earlier studies, Lancet realised that once you know the profile for a specific smell – once you know which receptors fire in response to that odour – you can dispense with that smell's actual molecular structure altogether. So now, instead of trying to recreate the identical scent through the whiffer device, Harel and Lancet are trying to recreate that scent's firing pattern in the olfactory epithelium, which would recreate the scent's effect on the brain. The strategy's only flaw: no one yet knows exactly how odours are registered in the brain. 'We don't understand the secret language of receptors,' says Lancet. 'If we did, we could generate the ultimate algorithm.'

To get round the problem, the scientists are using what they do know about olfaction to arrive at an approximation of what they don't know. What they do know is the profile from the electronic nose, which is easily obtained by exposing the sniffer to the selected scent. The human nose profile, however, is more difficult to determine, since there's no efficient and accurate way to record the firing patterns of nerve cells in the human nose.

However, it is possible to get indirect assessments in two ways. The first is by studying the olfactory receptor genes and proteins and gleaning information about how each receptor might bind different odorants. The second is by asking people to describe what they smell. So Harel and Lancet are putting together panels of volunteers and rubbing their noses in a variety of basic scents. They ask the participants to describe the smell and how closely it resembles other smells. The team then combines this information with the results from the e-nose and lets the algorithm loose.

The algorithm runs through the electronic and biological profiles to come up with a pattern that matches as closely as possible that of the desired odour. Phrased in an oversimplified equation, the process works like this: the e-nose profile + the profile from the human panel = the profile that tells the whiffer how to mix the odours.

Here's how that equation might work in practice. Let's assume that the electronic nose has two dozen different sensors. The exact firing pattern of these sensors constitutes the e-nose profile for a specific odour – say, pepperoni pizza. Now ask a group of volunteers to smell the mixture of various ingredients that make up an actual pepperoni pizza and create a profile based on their responses. Then mathematically combine the e-nose profile with the human panel profile to arrive at an approximation of the original scent. The molecular structures of the resulting computer-generated odorant mixture may not be identical to the original smell, but it will elicit the same firing pattern of receptors in the olfactory epithelium; and as far as the brain is concerned, that's enough – the rhinencephalon won't be able to tell the difference.

Harel and Lancet would like one day to put this technology into little set-top boxes that they claim will make everything from digital kodo to odoriferous game shows a reality. 'Machines that can manufacture smells will soon enter the home and office,' Lancet says, 'because people want to do digitally everything that they do naturally. People want all that sensory titillation at home.'

Or do they? Some of the companies promising to flood the home with fragrance through aromas transmitted over the Internet have recently gone bust. As the dot-com bubble burst, plans to scent-enable the e-commerce sites of perfumiers, florists, cosmetics companies, prepared-food corporations and cigar manufacturers have fizzled out. A more likely first application of this technology is in computer games. In a video game enhanced with scent technology, you'll actually feel the sting of gunpowder in your sinuses after you blow away some mutant alien villain. The goal is to make these kinds of games even more immersive and compelling by adding odours such as gunsmoke and roadkill, which is what thirteen-year-old boys want to smell.

If scent-enabled websites eventually do become a reality, Paco Underhill, a retail anthropologist and author of the book *Why We Buy: The Science of Shopping*, might finally get an answer to his question, 'Can you smell a ripe peach online?' The question is not a Zen koan for New Age netizens, but a serious challenge to the usefulness of scent-enabled websites.

Underhill, CEO of Envirosell Inc., a research and consulting firm, is underwhelmed by all the fuss about devices like Aerome's MediaScenter. He thinks scented websites will have only limited consumer appeal. 'Buying a fragrance is not about how it smells,' he says, 'but how it smells on *you*. Fragrances smell different on different people, and pumping them out of a computer just turns smell into a room deodorant.'

According to Underhill, shopping 'involves experiencing that portion of the world that has been deemed for sale, using our senses ... as the basis for choosing this or rejecting that. It's the sensory aspect of the decision-making process that's most intriguing because ... virtually all unplanned purchases – and many planned ones, too – come as a result of the shopper seeing, touching, smelling or tasting something that promises pleasure, if not total fulfilment.'

Sound and vision have always been our primary interfaces with the worlds inside computers. Smell was long consigned to second-class sensory status; odours were meant to be suppressed rather than expressed. As companies like Aerome put odours online, will shoppers follow their noses onto the Web? Will scent technology bring some of the wild, natural smellscapes of the world to our desktops and televisions, or simply become yet another intrusive way to deliver an advertising message?

The technology is still too immature and the applications still too limited for a definitive answer. We might find, like Montaigne, that the invention of odours will 'render us more capable of contemplation.' Or we might well wish that everything would just stop making scents.

FIVE

TASTE

Fun with Electronic Tongues

'De gustibus non est disputandum.'

The Futurists had strange table manners. Led by Italian artist Filippo Tommaso Marinetti, the members of this early twentieth century avant-garde movement eschewed cutlery, preferring to bury their faces in a dish so as better to savour its fragrance and flavour. Gathering at the Holy Palate Restaurant, they would sup on such exotic entrées as Sunshine Soup, Sculpted Meat, Elasticake and Cubist Vegetable Patch, a selection of celery and carrots cut into squares and sprinkled with paprika and horseradish.

In *The Futurist Cookbook* Marinetti, a poet, painter and polemicist, proclaimed that the 'Futurist culinary revolution ... has the lofty, noble and universally expedient aim of changing radically the eating habits of our race, strengthening it, dynamizing it and spiritualizing it with brand-new food combinations in which experiment, intelligence and imagination will economically take the place of quantity, banality, repetition and expense.'

The Futurists' intelligence and imagination were lavished on a series of bizarre and elaborate dinners in which 'every person has the sensation of eating not just good food but also works of art'. The cookbook offers recipes for every occasion. During the 'geographic dinner', for example, guests select their meal by pointing to a spot on a map of Africa painted onto the waitress's dress. For those hankering after an airline meal, the 'aeropoetic futurist dinner' takes place in the cockpit of a plane flying at 3000 metres. And for those on a low budget, the 'economical dinner' features 'apples cooked in the oven, then stuffed with beans which have been boiled in a sea of milk'.

Perhaps the most astonishing evening ever conceived by the Futurists was the 'tactile dinner party'. According to Marinetti's meticulous instructions, the 'host has carefully prepared ... as many pairs of pyjamas as there are guests: each pair of pyjamas is made of or covered with a different tactile material such as sponge, cork, sandpaper, felt, aluminium sheeting, bristles, steel wool, cardboard, silk, velvet, etc. A few minutes before the dinner each guest must put on, in private, one of the pairs of pyjamas.'

After the guests are suitably attired, they are led into a dark room where they must choose their dinner partner by the texture of his or her pyjamas. Once the choices have been made, everyone enters the dining room and the meal commences. The courses are brought in – Polyrhythmic Salad and Magic Food are just two – and every time a diner lifts his head from the plate (remember, no utensils) a waiter squirts perfume into his face. Between one dish and the next, Marinetti insists that 'the guests must let their fingertips feast uninterruptedly on their neighbour's pyjamas'.

Though the Futurists' ideas about food were outlandish and often impractical, their fantastic recipes celebrated the mystery of taste. About fourteen years ago, Kiyoshi Toko became similarly possessed, albeit during a far less exotic meal. While lunching in the cafeteria at Kyushu University in Fukuoka, Japan, Toko, a professor in the department of electronics, overheard a couple at the next table arguing about the coffee. One said it was bitter; the other said it was not.

Toko was puzzled. These two people were drinking the same coffee poured from the same pot. How could there be any disagreement about the taste when the chemical composition of both cups was identical? The dispute inspired Toko to pursue a course of research that eventually led to the development of the world's first electronic tongue, a device that in many ways can indeed account for taste.

GUESS WHO'S COMING TO DINNER

Like smell, taste is notoriously difficult to explain. The two senses are, in fact, intimately related. More than 90 per cent of a meal's flavour – apart from the four basic tastes of sweet, sour, bitter and salty – is actually fragrance, which rises up from food during chewing and is forced across the olfactory epithelium through the nasopharynx at the back of the throat.

Taste and smell most likely emerged from a single precursor sense, some kind of rudimentary chemical ability that allowed bacteria and other organisms to discriminate between potential nourishment and poison. French judge and gourmand Jean-Anthelme Brillat-Savarin noted this sensual mix in *The Physiology of Taste*, published in 1825. 'I am not only convinced that without the cooperation of smell there can be no complete degustation,' he wrote, 'but I am also tempted to believe that smell and taste are in fact but a single sense, whose laboratory is the mouth and whose chimney is the nose; or to be more precise, in which the mouth performs the degustation of tactile bodies, and the nose the degustation of gases.'

Anyone in doubt about the link between smell and taste should talk to Colin Berry, an English molecular biologist who up until 1985 could smell perfectly. In that year he suffered a skull fracture during a rock-climbing accident that severed the nerves from his olfactory epithelium to his brain. Berry now has anosmia: no sense of smell at all.

'Anosmia has destroyed my sense of taste,' he says. 'I can still sense sweet, sour, bitter and salty. But flavour, which is almost all smell, is totally gone.' Berry can't tell when milk has turned sour, he can't enjoy lemon meringue pie, he can't savour fine wine. He eats spicy foods as often as possible, since these are the only meals that still generate some sensation on his tongue.

In anosmia, it's the inability to smell that deprives a person of their sense of taste. So Berry's tongue is just like everyone else's: a lunar landscape pitted with around 8000 to 10,000 taste buds, each of which is equipped with between fifty and seventy-five chemical taste receptors. The receptors, which expire and are replaced every ten days, are mostly clustered along the back, sides and tip of the tongue, though some cleave to the palate and others are scattered as far back as the throat. These specialised cells function very much like the receptors in the olfactory epithelium: they bind to taste molecules, translate those molecules' chemical signatures into neurotransmitters, which stimulate nerve fibres to fire off electrical signals to the brain, which the brain then interprets as a flavour.

Studies at the Monell Chemical Senses Center in Philadelphia suggest that several receptors for the neurotransmitter glutamate may be associated with a possible fifth basic taste, *umami*, a Japanese word meaning 'voluptuous savouriness'. Appropriately enough, foods containing lots of monosodium glutamate tend to have the *umami* taste. So the next time you sit down to a plate of Chinese noodles be prepared for a cascade of glutamate in your brain. But like smell, just how the brain pairs these specific molecules with that specific flavour – and which neurotransmitters correspond with which other tastes – is still very much a mystery.

The biological mystery of taste is made even more impenetrable by the sense's utter subjectivity. There's usually little argument about whether something looks black or white, sounds loud or soft, feels hot or cold. But as Toko's lunch anecdote shows, reasonable people can disagree about the bitterness of coffee. Which may not be surprising, since sight, hearing and touch respond to a single physical stimulus – light waves, sound waves and tactile pressure, respectively – while taste responds simultaneously to an enormous range of chemicals.

The average cup of coffee, for example, contains more than 1000 different chemical components, none of which is tasted in isolation but only as part of the overall flavour. Because different tongues respond more or less strongly to different flavours, one man's java may be another man's poison. 'Taste is very obscure,' Toko says. 'I wanted to develop an objective scale of taste, so that discussions about the bitterness of coffee could be more scientific and reliable.'

Developing this objective scale was made easier by the fact that there are just five basic tastes. Since it's not practical – or even possible – to make a sensor for every potential taste molecule, Toko set out to make sensors that mimicked the human taste receptor's ability to respond to one or more of these five basic tastes. By eliminating smell from the equation, Toko came up with a device that could detect the general taste of foodstuffs and beverages by their sweet, sour, bitter, salty and *umami* characteristics alone. 'If we can discriminate the taste from the smell,' Toko says, 'we can discuss flavour quantitatively.'

Though Toko sought to replicate the function of the human tongue, his electronic version looks absolutely nothing like the real thing. About the size of a very large blender, Toko's e-tongue some-

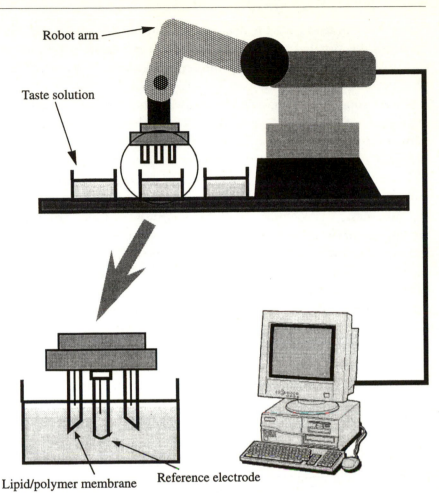

Figure 6. Toko's electronic tongue, as manufactured by the Anritsu Corporation.

what resembles the chamber of a revolver – an array of cylindrical plastic tubes arranged in a circle and serviced by a robotic arm. The tubes contain the substance to be sampled – anything from mashed potatoes to mineral water – first rendered, if necessary, into liquid form.

The robotic arm has seven appendages, each one bristling with lipid/polymer membranes that are fitted with electrodes. The arm first dips its gustatory digits into a saline solution that is used as a baseline reference. Then it sips the contents of a sample tube that contains, say, mashed potatoes. The sensors 'taste' the contents of both tubes and feed back the data to a computer. The computer then measures the voltages from the electrodes in the saline solution and the mashed potatoes, and the difference between the two is the numerical equivalent of the taste.

Toko's tongue wouldn't be much use without some kind of reference point for each of the five basic tastes. If the baseline for 'average saltiness' is 100, for example, and the batch of mashed potatoes measures 200, we know that these are very salty spuds indeed. So Toko took one substance as a reference guide for each of the five tastes: sodium chloride for saltiness, quinine for bitterness, sugar for sweetness, hydrochloric acid for sourness and the voluptuous savouriness of monosodium glutamate for *umami*.

Since each membrane responds to more than one taste – one membrane may have a weakness for sweetness and *umami* but not react at all to saltiness, while another membrane may react strongly to saltiness but not react at all to *umami* – each substance also generates a unique pattern of electrical responses across the electrode array. Different patterns mean different tastes. If a new brand of candy is being tested, for example, the results from the electronic tongue can be compared to those for the sugar reference to determine just how sweet it is.

Not content to let his device remain a laboratory curiosity, Toko set out to prove its discriminating palate on one of the human tongue's most demanding tasks: taste-testing beer. Toko took his tongue on a binge to classify eight different brands on an objective scale of sharp and mild tastes. The machine was easily able to distinguish differences among all the samples, with US brew Coors coming in at the top of the sharp, light end of the scale and Ireland's famed Guinness leading the rich, mild end of the spectrum. Unfortunately, Toko did not report on any of the experiment's side effects, so we don't know if his device experienced the dry mouth and sandpaper tongue common among heavy beer-drinkers. Toko also used the device to monitor the fermentation and blending of the special types of rice used to make sake. The taste sensor functions as an acidity meter while detecting changes in alcohol content during the fermentation process.

In Japan, the Anritsu Corporation has commercialised a taste sensor based on Toko's design. The device is used mainly by companies in the foodstuffs, pharmaceutical and packaging industries to monitor taste on production lines. Drug manufacturers also use the sensor to test the taste of new medicines.

In *Charlie and the Chocolate Factory* by Roald Dahl, Mr Wonka demonstrates a wonderful device he's invented for delivering chocolate bars directly to homes: Television Chocolate. After providing a quick overview of how ordinary television operates, he goes on to describe how he came upon the inspiration for his new device: 'If these people [television producers] can break up a *photograph* into millions of pieces and send the pieces whizzing through the air and then put them together again at the other end,' he asks, 'why can't I do the same thing with a bar of chocolate? Why can't *I* send a real bar of chocolate whizzing through the air in tiny pieces and then put the pieces together at the other end, all ready to be eaten?'

To the astonishment of his audience, Mr Wonka then demonstrates the device. A team of Oompa-Loompas clad in protective space suits teleports a bar of chocolate from one end of the room into a television set at the other, from whence the book's hero Charlie Bucket, plucks and eats it. 'Just imagine,' Mr Wonka cries, 'when I start using this ... you'll be sitting at home watching television and suddenly a commercial will flash on to the screen and a voice will say, "Eat Wonka's chocolates! They're the best in the world! If you don't believe us, try one for yourself – *Now*!" And you simply reach out and take one! How about that, eh?'

TriSenx, a Savannah, Georgia-based firm, is using a more technically feasible version of Willie Wonka's idea to deliver fast food over the Internet, which means that diehard geeks might one day never have to leave their desktops – not even to order pizza. Using technology similar to the MediaScenter described in the chapter on 'Smell', the company's UltraSenx machine works like a kind of gustatory fax: it transmits a message to the user's computer in response to a click on a taste-enabled website. From this message, a kind of miniature kitchen attached to the user's computer then whips up the appropriate flavour. 'Television and computers only provide stimuli for sight and hearing, leaving taste to the imagination,' says Ellwood Ivey, president of TriSenx. 'UltraSenx adds taste to sight and sound for a more complete sensory experience.'

If you're perusing the online menu of an Indian restaurant, for example, and want to sample the chicken curry before placing an order, just click on that item and the UltraSenx sends the recipe to a flavour printer, or 'sensualizer' as TriSenx calls it, about the size of a hat box. The printer contains a larder of basic flavours in liquid form that can be combined to create a variety of tastes, such as chicken curry. Once the printer gets the recipe from the UltraSenx it deposits the corresponding melange of flavour droplets onto a thin piece of edible rice paper and prints it out. The wafer is then ejected from the device and is ready to eat. Web cookies will never be the same again!

Computers can now not only prepare a meal for you; they can also feed it to you. Secom, a Japanese firm involved in the security, health care and insurance industries, has developed a feeding robot called 'myspoon' designed as an eating assistant for quadriplegics. The device — a robotic arm with a spoon and a fork for fingers that are agile enough to pick up rice and tofu — grasps pieces of food and brings them to the user's mouth. The user controls the arm by means of a laser pointer worn on the head that activates images of various foodstuffs on a computer display. Point the laser beam to an image of grilled eel, for example, and illuminate 'eat' and the robot arm swoops down with fork and spoon to grab it for you.

Toko even imagines the day when miniature taste-sensor technology will be attached to the ends of chopsticks and spoons. Dip your chopsticks into a meal and they will not only tell you what you are eating, but also list the ingredients and provide you with the recipe. Back home in your Internet-enabled kitchen, just plug the chopsticks into the fridge and the fridge will call up the recipe on the screen and order any missing ingredients. 'We are at the beginning of a new age of food culture,' Toko says. If so, the electronic tongue could well spark a whole new wave of Futurist dinners. Don't forget your pyjamas.

A LITTLE TASTE OF TEXAS

Though fast food over the Net may take some time to catch on, the days of Toko's hand-held tongue may be here thanks to researchers at the University of Texas at Austin. They've developed an electronic tongue that is as portable as a thermometer, thanks to an ingenious combination of microelectromechanical systems and chemistry.

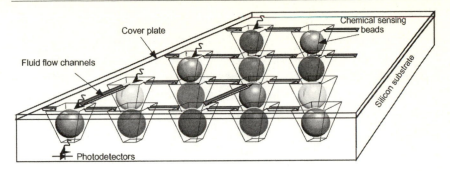

Figure 7. The Texas electronic tongue, showing the micromachined bead array and fluid flow channels for analysis of liquids.

The Texas tongue consists of artificial taste buds etched into a silicon wafer, or sensor chip. Inside each taste bud cavity is a tiny polyethylene glycol and polystyrene bead, to which one of four different chemical sensors is attached. There are hundreds of these beads, each of which sits in its own little concavity, like marbles in the Japanese board game Go. The beads act as taste receptors: when a liquid washes over them, they soak up the solution like a sponge and change colour in response to the presence of sweet, sour, salty or bitter tastes. One sensor, for example, might turn yellow when exposed to an acidic substance, while another might turn purple in response to something sweet.

The resulting colour spectrum is captured by a digital camera and transmitted to a computer, where a neural network analyses the chemical composition of the substance based on the colour changes in the beads. The computer assigns each solution its own distinctive 'taste', which is stored in a database for future reference in identifying the same or similar flavours. The neural network allows the device to learn to detect and remember a wide range of novel taste combinations.

Though the Texas tongue has the silicon equivalents of taste buds and receptors, it doesn't work at all like a human tongue. Human taste receptors, like those for smell, are non-specific; they respond to a variety of different taste molecules. The bead receptors, on the other hand, have pretty boring palates – each one responds only to a single molecule. So it's the pattern of these colours as a whole rather than the individual colours themselves that identifies the taste.

Dean Neikirk, a professor of electrical and computer engineering at the university, who developed the technology with chemists Eric Anslyn, John McDevitt and Jason Shear, explains how it's done: 'The chemists make a beaker of one type of taste bud that responds to one type of chemical agent. Then they make another that responds to another type of chemical agent, and so on until the array has enough different artificial receptors to enable it to respond to broad classes of mixtures.'

This versatility makes the artificial tongue ideal for analysing solutions that contain a wide range of biological and nonbiological chemicals or, in Neikirk's words, 'for tasting things that we don't want to taste'. One commercial application the university hopes to develop is medical diagnostics. Give the tongue some blood or urine to drink and it can produce a quick rundown of any toxins or drugs present in the sample.

In the part of Texas where Neikirk lives, many people still draw drinking water from their own wells. The detection of pollutants in water is a tedious and time-consuming process because of the sheer number of substances that need to be analysed. The taste sensor could speed up the process considerably, since it responds to many types of substances simultaneously rather than to just a single contaminant at a time. Neikirk and company believe the tongue will be able to measure low concentrations of a whole range of pollutants within minutes. The device could also be used to check the authenticity of whiskies and vodkas, to taste the difference between cheap knock-off versions of soft drinks and the real thing, or even to nibble raw oysters to see if they're okay to eat.

This ability to discriminate quickly between a large number of substances in a complex mixture is one of the Texas tongue's main attractions. And it's good for tasting wine as well as water. Though it's not possible to know in precise detail every ingredient that goes into a wine, Eric Anslyn argues that the big picture — the overall combinations of pectins and tannins that combine to make a good wine — is knowable. 'When you press a bunch of grapes to make wine, you create roughly 200 chemical structures,' Anslyn says. 'No one, not even the wine-maker, knows each and every chemical structure that goes into the wine. The only way to understand a mishmash of stuff like that is to throw another mishmash at it.'

The Texas taste receptor beads are that mishmash. The various chemical agents in the wine create a unique pattern in the beads that becomes diagnostic for that particular mixture. Human sommeliers can then taste the wine and if one blend is particularly successful, the wine-maker can use the e-tongue's recipe to recreate it. 'You still don't know each and every pectin and tannin,' says Anslyn, 'but the tongue can tell you that this particular mishmash of pectins and tannins makes a good wine and that mishmash makes a bad one. You know if you've got a winner.'

THE REVOLVING DOORS OF PERCEPTION

As Neikirk and Anslyn build the ultimate diagnostic mishmash, there's one biological mishmash that remains unresolved: synaesthesia, an unusual and little studied condition in which a stimulus received in one sense organ causes a sensation in another. 'It's kind of like figuring out that you have a belly-button,' says Karen Chenausky, a speech researcher who lives in Somerville, Massachusetts. 'At some point you just notice and start playing with it. Then you get really into it. And after a while you get bored, because you realise everyone has one. Except not everyone has this.'

Chenausky is talking about one of the most common forms of synaesthesia: coloured hearing. For her, sound and vision mingle: the different tones of words and letters involuntarily evoke distinct and vivid colours in her mind. Russian novelist Vladimir Nabokov was similarly gifted. In his memoir *Speak, Memory*, he lovingly recites his own private alphabetic palette of sounds: 'In the green group, there are alder-leaf f, the unripe apple of p, and pistachio t. In the brown group, there are the rich rubbery tones of soft g, paler j, and the drab shoelace of h.

In his poem 'Vowels', French poet Arthur Rimbaud recited a more sinister spectrum of coloured sound:

E, candour of sand and pavilions,
High glacial spears, white kings, trembling Queen Anne's lace;
I, bloody spittle, laughter dribbling from a face
In wild denial or in anger, vermilions.

Nikola Tesla, the eccentric inventor of electrical devices, had a particularly bizarre form of synaesthesia. 'When I drop little squares of paper in a dish filled with liquid,' he wrote in his collection of autobiographical essays, *My Inventions*, published in 1918, 'I always sense a peculiar and awful taste in my mouth.' But like Nabokov, Chenausky revels in this added perceptual dimension. 'Synaesthesia is an extra way of perceiving the world,' she says. 'The parts of the world I perceive in this way are parts I hold most dear.'

Though known for roughly the past 300 years, synaesthesia – derived from the Greek words *syn* (together) and *aisthesis* (to perceive) – is still unexplained. Scientists don't agree on what causes it, or even how widespread it is. According to neurologist Richard Cytowic, author of *The Man Who Tasted Shapes* and *Synesthesia: A Union of the Senses*, only ten individuals in a million are synaesthetes. Other surveys suggest that about one in every 2000 people automatically sees colours when hearing words, letters or numbers. All researchers agree, however, that the vast majority of synaesthetes (roughly 90 per cent) are female and that by far the most common form of synaesthesia is coloured hearing of the kind that Nabokov and Rimbaud describe.

Some investigators dismiss synaesthesia simply as a learned association embedded in memory from childhood. According to this view, coloured hearing is just a case of a person remembering the colours of the alphabet books or toy letters she played with as a child. This does not explain, however, why synaesthetic twins and siblings almost never share the same colour–letter correspondences, a similarity that would be expected since presumably the same books and toys were used by the whole family. My wife and her mother, for example, both have coloured hearing, something I didn't discover until I started writing this book, but the colours of their alphabets are completely different.

The so-called 'sensory leakage theory' offers a biological explanation, suggesting that synaesthesia is the result of neurological crosstalk between the auditory and visual pathways in the brain. Because these two brain centres are anatomically close, it's possible that stray visual neurons, for example, could branch out into the neighbouring auditory region to transmit visual signals in response to sounds.

Cytowic, however, claims to have tracked the source of the synaesthetic experience back to the limbic system, one of the oldest parts of the brain and the site at which emotions and memories are processed. Cytowic calls synaesthetes 'living cognitive fossils' because he believes this kind of multi-sensory perception is as ancient as the place from which it originates.

According to his theory, synaesthesia may well have been our primeval way of experiencing the world, until the more rational cortex evolved and filed the senses away into their individual compartments. 'Synaesthesia is a normal brain function in every one of us,' Cytowic says, 'but its workings only reach consciousness in a handful. It may well be a memory of how early mammals saw, heard, smelled, tasted and touched.'

Other research indicates that synaesthesia may be due to a profusion of neural connections between the parts of the brain that control the five senses. Daphne Maurer of McMaster University in Ontario even suggests that this embarrassment of neurological riches makes all babies born synaesthetes who experience a rattle, for example, not just as an intriguing sound but as a barrage of colours, smells, tastes and tactile sensations. Only after about the age of four months, when the infant's cortex has sufficiently matured, does the synaesthesia fade. 'The brains of young babies have many more neural connections than they do in later life,' explains Simon Baron-Cohen, a lecturer in the Department of Experimental Psychology at Cambridge University who is studying the theory. 'Many of the links gradually get pruned back. Synaesthetes may be people who retain these neural connections.'

In some cultures, synaesthesia is regarded as a normal – and, indeed, essential – ability rather than some neurological aberration. The Dogon people of Mali, for example, routinely hear smells and smell sounds. The Dogon believe that the meanings of sounds and smells are fundamentally related because they are both communicated through the air. Thus, the sound of an eloquent speaker's voice is perceived as sweet-smelling, while the chatter of someone who makes lots of grammatical mistakes is accompanied by an unpleasant odour.

'Synaesthesia is not a disease,' Cytowic concludes, 'but a bonus. Your senses give you more than you bargained for.' Though still poorly understood, the experiences of Karen Chenausky and the Dogon suggest that there is more to the world than meets the eye, ear, nose, tongue and skin.

TAPPING INTO THE TONGUE

Robert Bradley, a professor of dentistry and physiology at the University of Michigan, wants to know what happens on the tips of our tongues. So he's tapping into the rat tongue to find out.

Bradley's device – a sieve electrode, produced in collaboration with Khalil Najafi – consists of a clutch of cuff electrodes that wrap around the nerve fibres like an arm slipping into a sleeve. The electrodes are inserted into a series of microscopic holes punched into a silicon disk about 4 millimetres in diameter, with each electrode linked to a miniature ribbon cable. These cables are then hooked up to an external connector, which is in turn attached to external amplifiers and analysis equipment. The rat's glossopharyngeal nerve, which is located at the back of the tongue and transmits gustatory information to the brain, is then severed and the sieve electrode is implanted on top of the nerve. Over a period of about ten weeks, the nerve regenerates through the micromachined holes and reconnects to the taste receptors on the tongue. Once the sieve settles into place, the computer records the nerve signals coming from the rat's taste buds.

Bradley's goal in inventing this unwieldy device is to explore the reasons behind the high turnover of tastebud cells. 'It's like watching ageing because the cells are born and die all within ten days,' he says. 'We want to know how the neural coding for taste works, to discover how we perceive this aspect of the external world, how it is encoded and processed.'

If taste buds are really non-specific – if each taste bud can actually respond to all tastes – then Bradley's tongue-tap could help prove it. 'There are no separate receptors in the eye for trees and flowers,' Bradley says. 'Vision is the result of a pattern of signals across the optic nerve. So taste might just be a pattern of signals across all the nerve fibres on the tongue.' The only way to investigate this is to record from multiple nerve fibres at the same time.

That's why Bradley is now working with the chorda tympani, the nerve that funnels information from all the little red spots on the tip of your tongue. It's a smaller and more difficult nerve to tap into than the glossopharyngeal nerve, but it feeds more taste receptors and so offers the opportunity to get more and better data, hopefully from at least thirty-two different nerve fibres. 'We need to look at multiple nerves to discover if there is a unique pattern for each taste and to be able to discriminate the nuances among tastes,' he says. 'Right now we don't know how the tongue tells the difference between salty and sweet, much less how it discriminates among the sweetnesses of chocolate, ice cream and rice pudding.'

Though Bradley's work is still in the early stages of development, the basic technology behind his device is being used as a model for the implantation of miniature electronics in other parts of the human nervous system. According to Bradley, similar interfaces between machines and the nervous system could be used to stimulate movement in victims of paralysis or create artificial hands powered directly by nerve impulses.

'If you can tap into a nerve,' says Bradley, 'you can feed information into and out of the central nervous system to power motors or microphones or whatever. The device potentially could be used as a neural interface to connect the body to anything.' Such an interface would put people in very close contact with computers; for some, perhaps too close for comfort. This type of man–machine interface is explored in the next chapter: Touch.

SIX

TOUCH
The World Is Your Interface

'In the electric age we wear all mankind as our skin.' MARSHALL MCLUHAN

When the organisers of the Fifth Australian Sculpture Triennale in Melbourne asked Australian performance artist Stelarc to submit a proposal for a site-specific installation, he came up with a novel location in which to situate the work: his stomach.

After fasting for eight hours, Stelarc (with the help of an endoscopist) ingested a 15mm × 5cm capsule made of titanium, stainless steel, silver and gold. The capsule was tethered to a flexidrive cable, which was linked to a control box outside his body. The stomach had to be inflated with air and drained of excess fluid with a pump before this metallic lozenge, jointly designed by a jeweller and a microsurgery instrument-maker, slithered its way into Stelarc's abdomen.

While Stelarc's stomach was thus distended, the capsule – now beeping and emitting flashes of light – unfurled to its full dimensions of 5cm × 7cm by means of a worm screw operated via a servomotor and logic circuit in the control box. It took six insertions of the device over two days to adequately document the performance using video endoscopy equipment, though for safety reasons it was impossible to film the fully extended sculpture inside the body. Only glimpses of the control cable and the head of the sculpture were captured.

'The intention was to design a sculpture for a distended stomach,' Stelarc says of the 1993 performance. 'The technology invades and functions within the body not as a prosthetic replacement, but as an aesthetic adornment. The hollow body becomes a host, not for a self or a soul, but for a sculpture.'

Stelarc is a living example of the strange and surprising ways technology is getting onto, and under, our skin. Computers are moving off the desktop and into everyday objects – and human bodies – putting people in touch with technology in an ever closer union.

Physicians are implanting electrodes into the bodies of patients to rehabilitate atrophied muscles, prevent epileptic seizures and restore motor function lost as a result of paralysis. Engineers are creating hybrid prosthetics such as ankles, legs and knees in which silicon chips are melded with living tissue. Computer scientists are designing haptic (from the Greek word meaning 'to touch') interfaces that allow users to reach out and touch digital information, transforming the plain old graphical user interface into a graspable user interface. By coupling digital information with everyday objects such as table tops, appliances and coffee cups, the physical world is becoming one enormous interface.

'Technology demonstrates the biological inadequacy of the body,' Stelarc says. 'We can't continue to merely design technology for the body because that technology begins to usurp and outperform the body. Perhaps it's now time to design the body to match its machines.'

Stelarc's work shows, albeit in an extreme form, what happens when metal meets meat, when the body merges with its machines. During one 1998 performance, Stelarc was connected to and carried by an elaborate high-tech exoskeleton: a six-legged, pneumatically powered walking machine able to move forwards, backwards and sideways, as well as pivot, squat and lift. Stelarc was positioned like a mechanical spider on a rotating turntable at the centre of the device, his upper body and arms wrapped in a jointed mechanism with magnetic sensors on each segment that allowed the control system to know the position and orientation of his limbs. He was thus able to operate and navigate the exoskeleton by moving his arms, his gestures choreographing the machine's motion.

For another event Stelarc wired himself up to the Internet. His body was dotted with electrodes – on his deltoids, biceps, flexors, hamstrings and calf muscles – that delivered gentle electric shocks (15–40 Volts), just enough to nudge the muscles into involuntary contractions. The electrodes were connected to a computer, which was in turn linked via the Internet to computers in Paris, Helsinki

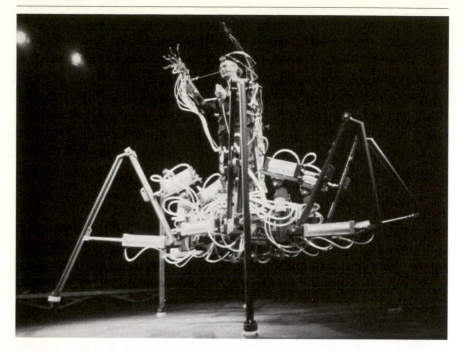

Figure 8. Stelarc inside his exoskeleton.

and Amsterdam. By pressing various parts of a colour-coded three-dimensional rendering of a human body on a touch screen, participants at all three sites could make Stelarc do whatever they wished.

The audience controlled Stelarc's movements just as they would operate an ATM machine. To get him to lift his leg and walk, they simply pressed the calf muscle on the touch screen and the computer sent a quick jolt to that set of electrodes – and the leg ascended while Stelarc maintained his balance. To get him to raise his arm, they just pressed on a bicep and the arm began to rise. As well as initiating movement by pressing the muscle sites, participants could compose more complex motions by pasting together sequences from a library of gesture icons and pressing 'play'.

Stelarc's body did not move in a jerky way, because the voltage was delivered incrementally. He felt a slight tingling sensation that gradually grew stronger as he watched his fingers curl or his biceps contract. His limbs bent and swung smoothly. As the jolts squirted through his skin, Stelarc says he felt a kind of 'split physiology, intimacy without proximity. A part of your body becomes alien, moving without your agency but through some remote body over the Internet.'

This kind of split physiology, Stelarc believes, could soon make its way from performance art venues into living rooms, kitchens and bedrooms:

> Just as the Internet provides interactive ways of displaying, linking and retrieving information, it may now allow unexpected ways of accessing the body itself. The Internet could become a kind of crude external nervous system, reconstructed not as a medium for the transmission of information but the transmission of physical acts. Imagine the possibilities: dancers or sexual partners, remotely situated, each able to control half of the other person's body. It would be much more powerful than simply seeing or speaking to someone over the Net. It would be cosmetic surgery for the senses.

Stelarc – an intense, powerfully built man who began his artistic career in the late 1960s as 'a bad painter' and decided to do performance art instead – speculates about one possible future in which human beings might resemble benign versions of the Borg, the malevolent alien race on *Star Trek: The Next Generation* that is joined in a group mind thanks to implanted electronics. In this scenario, though, individuals are able to seamlessly slide between individual and extended operational systems, to perform both beyond current biological limits and outside the normal constraints of space. Stelarc would like to see the Internet transformed from a means of information transmission to a mode of effecting physical action, so that the body becomes both an object of desire and an object to be redesigned and rewired. As Stelarc himself says:

> Prosthetics have always been a sign of lack – the loss of a limb or the loss of some motor function. But prosthetics could become a sign of excess and enhancement. We have to turbo-drive the body and connect it to the Net. At the moment this is done indirectly via keyboards and other devices. There's no way of directly jacking in. What will be interesting is when we can miniaturise these technologies and implant them directly into the body.
> We shouldn't have a Frankenstein-like fear of incorporating tech-

nology into the body and we shouldn't consider our relationship to technology in a Faustian way — that we're somehow selling our souls because we're using these forbidden energies — nor should we be obsessed with control and constraint. Technology is, and always has been, an appendage of the body.

SURGERY FOR THE SENSES

What Stelarc is doing out of artistic choice, Brian Holgersen, a thirty-year-old Danish tetraplegic, is doing out of physical necessity. For Holgersen, technology as an appendage of his body is an everyday reality.

Eight years ago, just after graduating from university with a degree in civil engineering, Holgersen took a motorcycle trip to the UK to visit his sister. While in the UK he was in an accident and broke his neck. Except for some minor movement in his shoulders, left arm and left hand, he was paralysed below the neck. Six years ago Holgersen underwent an experimental surgical procedure to implant a neural prosthesis, an interface between an electronic device and the human nervous system. These types of implant can bypass damaged stretches of the spinal cord and are meant to restore or replace sensory and motor function in victims of paralysis or amputees.

Paralysis results from neck and spinal cord injuries because the neural traffic that moves between the brain and the muscles — one lane to the brain from the muscles and another to the muscles from the brain — is severed or blocked. Like a kink in a garden hose, spinal trauma cuts off the flow of information that travels along afferent nerves, which send signals from the body to the brain, and efferent nerves, which carry instructions from the brain to the body's musculature. As a result, information from the skin and muscles — about the weight, temperature and texture of objects, for example — never reaches the brain, and motor commands from the brain never reach the rest of the body. In many cases of paralysis, though, the motor and sensory nerves below the level of the lesion remain intact and could function again.

Functional electrical stimulation (FES), which employs advanced electronics and sensors to stimulate precisely controlled muscle con-

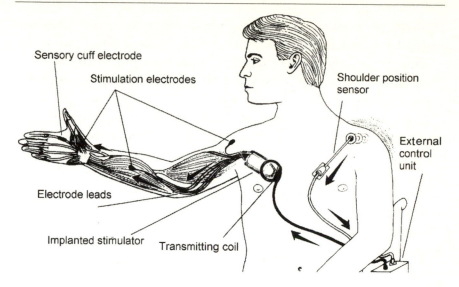

Figure 9. A schematic illustration of a system for the restoration of grip function in patients with spinal cord injuries.

tractions, can be used as a detour around spinal cord lesions. For the past several years, FES has been used to restore basic motor functions – such as grasping objects, moving the legs or controlling the bladder and bowels – to paralysed patients after severe spinal cord injury.

Holgersen uses the Freehand system, a device developed at Case Western Reserve University in Cleveland, Ohio, and manufactured by the NeuroControl corporation, that restores the ability to grasp, hold and release objects. During a seven-hour operation, surgeons made incisions in Holgersen's upper left arm, forearm and chest. Eight flexible cuff electrodes, each about the size of a small coin and similar to those used in the visual prosthetic described in the chapter on 'Sight', were attached to the muscles in his arm and hand that control grasping. These electrodes were then connected by ultra-thin wires to a stimulator – a kind of pacemaker for the nervous system – implanted in his chest. The stimulator was in turn linked to a position sensing unit attached to Holgersen's right shoulder, over which he retains some motor control. The whole apparatus is powered by an external transmitting coil worn above the location of the implanted stimulator.

When Holgersen wants to pick up a glass, he moves his right shoulder upwards. This movement sends an electrical signal from the position sensor, which is worn under his clothing, to the stimulator in his chest, which amplifies it and passes it along to the appropriate muscles in his arm and hand. In response, the muscles contract and his left hand closes. When he wants to release the glass, he moves his right shoulder downwards and his left hand opens.

'It's strange when you first use it,' Holgersen says of the device. 'I move my right shoulder and see my left hand move. But I quickly got used to it and now it feels very natural. I don't even think about it. It has become part of me and made me more independent.' Thanks to the Freehand implant, Holgersen can now hold a cup, lift a fork and grasp a pen, actions he was previously unable to perform.

The Freehand system works by using the electrodes to make Holgersen's muscles respond as if they were still receiving instructions from his brain. For Holgersen, though, the brain that controls his left hand sits on his shoulder and inside his chest. He has no sensation in the hand, but uses an experimental sensing system developed at Aalborg University in Denmark to gauge the weight and hardness of objects so that the grip can be relaxed or tightened accordingly.

Though use of the Freehand system comes naturally to him now and has increased his autonomy, Holgersen's grasp is still clumsy and he still requires round-the-clock care. Like most patients, he was able to master the device after about four weeks of practice. One study of the system's effectiveness found that seven out of nine Freehand users reported good independent function after about two to four weeks of training. The tasks they considered most important included writing, using a knife and fork, opening doors, brushing teeth, shaving and using the phone. 'With practice, it gets better,' he says, 'and makes daily things like eating and brushing your teeth much easier. People are usually not even aware that I'm using it. When they find out, they are astonished that it's even possible.'

The Freehand is not for everyone, though. To benefit from the device, patients must have use of a shoulder and upper arm and partial use of their hands. The technology can be fragile, too, and patients must be constantly on guard against infection around the implanted electronics. And since the Freehand provides no tactile feedback regarding temperature, users also have to be careful when handling hot objects such as cigarettes or coffee. Research is under way to make the implants smaller, lighter and more ergonomically and cosmetically acceptable. A wireless version of the system, which would make it more comfortable and even less conspicuous, is also under development.

Perhaps the biggest drawback to the Freehand system, however, is that it is limited to restoring motor control. The device does not restore sensation, so users cannot tell whether an object is hot or cold, soft or hard, smooth or rough. Thomas Sinkjaer and his colleagues at the Centre for Sensory-Motor Interaction at Aalborg University are developing neural prosthetics that can actually feel the texture of objects and transmit this information back to the user. 'We want to make patients aware of the parts of their bodies that they cannot sense,' says Sinkjaer, who has worked with Brian Holgersen for the past six years. 'We want to use the information transmitted by the natural receptors in the skin to get the brain to figure out where the stimuli are coming from.'

This kind of sensitive prosthetic would not only employ efferent nerves to send motor control instructions to the muscles, but would also recruit afferent nerves to send tactile information from paralysed limbs to other parts of the body, where the sensations could be perceived. With such a device Holgersen might feel the shape and weight of a freshly brewed cup of coffee he is holding, for example, as a tingling sensation on his cheek: the heavier the cup, the more intense the tingle.

The same technology could be used by amputees, too. Someone with an artificial hand, say, might sense the firmness of a handshake as pressure exerted on the back of the neck. In one experiment with a patient who had lost a hand, Sinkjaer's team discovered that stimulation of a nerve once connected to the thumb of the amputated hand evoked a sensation that seemed to emanate from the missing thumb. In other words, the nerve stimulation brought a phantom thumb to life. Prosthetic limbs could become much more efficient and realistic if nerve signals from a phantom thumb, for example, could be matched to an artificial thumb.

Making this kind of tactile feedback work – and making it feel real – will be difficult. Since afferent nerves are the messengers of sensation, their recruitment in these types of prosthetics is crucial to approximating the original tactile perceptions. Sinkjaer and his colleagues have experimented with afferents that evoke sensations of tapping on and poking at the skin; but hundreds, perhaps thousands, of different afferent nerves, including those that handle such perceptions as shape, texture and temperature as well as contact and release, would need to be stimulated to create convincing sensations. This is a daunting task, since the hand has the highest density of nerve receptors in the entire body, about 20,000. Each finger alone has about two thousand touch receptors.

Morten Haugland and Sinkjaer have designed and built a tactile feedback system controlled by means of signals from implanted cuff electrodes. A volunteer tetraplegic, who already has a Freehand implant, has a nerve cuff electrode implanted around the cutaneous nerve innervating the index finger. The goal of the project is to learn how to extract and use relevant information from the recorded nerve signals to allow the patient to better calibrate the force of his grasp.

The act of grasping and holding an object is not as straightforward as it seems, though. It's actually accomplished in fits and starts. When you lift an object like this book, you exert continuous force to hold it in place. This pressure is then strengthened or weakened depending on feedback from the hand and its various movements. So, in some respects, holding this book is a process of constantly letting it go and catching it again. This can only be accomplished thanks to the detailed tactile information fed to the brain by afferent nerves. Tetraplegics are deprived of this information, which is why grasping with the Freehand system can be a hit-and-miss affair.

Andreas Inmann, a graduate student at the Centre for Sensory-Motor Interaction, is trying to tap into this grasp-specific feedback so that Freehand users can maintain a steadier grip. When the grasp becomes too loose, the afferent nerves in the instrumented index finger fire more strongly. The electrodes detect this and the hand muscles receive an increased stimulation to prevent the object from falling. With more grasp-specific information, the stimulator could gradually increase or decrease the amount of stimulation needed to keep the object in place. This would also help minimise the amount of electrical stimulation delivered to the muscles, an important advantage since high levels of stimulation over prolonged periods can damage tissue.

Previous systems provided only very crude feedback information, so a Freehand user could never be sure whether the force exerted was too weak to hold a glass or strong enough to crush it. Inmann's system is capable of automatically regulating the grasp force to securely hold the object. His ultimate goal is to use this kind of information to create responsive prosthetics that warn users if an object is slipping away or if pressure on the prosthetic limb is excessive.

MONKEY BUSINESS

Richard Andersen, a neuroscientist at the California Institute of Technology in Pasadena, is also working on neural prosthetics for victims of upper spinal cord injuries and neurological disorders such as strokes. But instead of splicing into the nervous system at the hand or arm, he is going straight to one of its most important parts: the brain.

Andersen focuses on the posterior parietal cortex, the area of the primate brain that integrates visual and sensory-motor information. It is here that the brain coordinates what you see with what you do. Working with rhesus monkeys, Andersen and colleagues discovered an area within the posterior parietal cortex that encodes the animal's plans for its next intended arm movement. Andersen wants to decipher the brain signals from this region and use them to operate biological and prosthetic arms.

Though Andersen's work is on the cutting edge of neuroscience, in essence it is not unlike the experiments first conducted in 1786 by Italian physician and physicist Luigi Galvani. In that year Galvani made the legs of a dead frog twitch by touching its exposed nerves with a pair of scissors during an electrical storm. Galvani concluded from his experiments that living tissue contained an innate force that he called 'animal electricity', which was discharged by the brain and conducted by the nerves to the muscles.

Apart from his categorisation of animal electricity as a novel elemental force, Galvani was correct about how electrical impulses stimulated muscles to contract; and, indeed, the brain transmits as well as receives these electrical signals. What Galvani did not know, and what Andersen and others have discovered, is that during voluntary movements nerve cells in the brain start firing off electric signals well before any action actually takes place: the body lags slightly behind the brain's intention to act. It's as if the brain warms up for an impending movement by directing specific clusters of neurons to fire, just as before setting off a driver warms up a car by pumping the gas pedal. What Andersen pinpointed in the posterior parietal cortex was the spot at which the brain prepares to instruct the arm to make a move.

This little bit of brain physiology is crucial for the development of neural prosthetics because paralysed people are still able to plan the actions of their immobile limbs, but their spinal cord is unable to transmit the electrical signals necessary to make it happen. The brain is willing but the body is weak. 'We want to decipher this neural code so subjects can use it to move prosthetic limbs,' Andersen says.

To find out how this planning process is carried out, Andersen implanted electrode arrays in the part of the monkey posterior parietal cortex that plans arm movements, an area known as the parietal reach region (PRR). He then recorded neural activity here, which begins when the plan to reach is first formed and ends when the actual reaching motion is initiated.

Andersen and his team discovered that each neuron in the PRR responds to a particular target in space and may not be concerned with the fine details of the movement, which makes the PRR an ideal source for signals to control prosthetic limbs. 'The PRR is where decisions about arm movement are made in the non–human primate brain,' says Andersen. 'The monkey sees where it wants its hand to be and the PRR makes the plan to get it there.'

Andersen's goal is to understand the neurobiology of this process so that he can tap into the PRR signals, reroute them around damaged stretches of spinal cord and deliver them to a neurally-controlled prosthetic arm or a biological arm, in the latter case through functional electrical stimulation like that used by Brian Holgersen. Ultimately, users of such a device would be able to operate a prosthetic or virtual arm simply by mentally planning the movement. 'With this technology, people could surf the Internet and move their own bodies and artificial limbs by thought alone,' says Andersen. 'It's technically complicated, but I believe it is doable.'

In fact, part of it as already been done. By implanting electrodes in rats' brains, for example, researchers at MCP Hahnemann School of Medicine in Philadelphia have taught the animals to use a water-

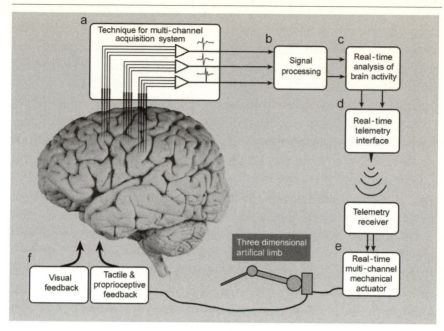

Figure 10. A schematic diagram showing how a robotic arm could be controlled using brain-derived signals. Microelectrode arrays sample brain activity. These data are transformed by a mathematical algorithm into trajectory signals that are used to control the movement of the robotic arm. The subject is provided with both tactile and visual feedback signals generated by the robotic arm.

dispensing robot just by thinking about it; and at Emory University in Atlanta, Georgia, brain implants given to disabled individuals enabled them to control a computer cursor simply by imagining the movement of various parts of their bodies. Miguel Nicolelis, associate professor of neurobiology at Duke University Medical Center in North Carolina, has taught monkeys to control a robotic arm via their brain signals. The best-laid plans of mice, men and monkeys are finally being deciphered.

Working with colleagues at Duke, MIT's Laboratory for Human and Machine Haptics (also known as the Touch Lab) and the State University of New York Health Science Center, Nicolelis implanted multi-electrode arrays – one with ninety-six electrodes and one with thirty-two electrodes – into the brains of two owl monkeys. The electrodes were cast across three different brain regions, including the PRR, each of which was involved in the planning and execution of arm movements. Nicolelis and colleagues monitored the monkeys' brain signals for two years as they carried out various tasks, like reaching for food, that required voluntary arm activity. Eventually the researchers were able to isolate and identify the signals that preceded the reaching movements.

Nicolelis and his team then routed the monkey's brain signals through a computer, which was able to decode in real time where the monkey was going to reach. As the monkey started to grasp for the food, the computer picked up the neural traffic and forwarded it to a robotic arm called the Phantom. When the monkey extended its arm, the Phantom precisely mimicked the action. Nicolelis even transmitted the brain signals over the Internet to the Touch Lab in Cambridge, Massachusetts, so the monkey's neural commands operated another Phantom located 965 kilometres away.

Nicolelis is convinced that this system will work for humans as well, forming the basis of a direct brain–machine interface. This interface might not only allow paralysed people to control their own biological or prosthetic limbs in real time, but could also open up the possibility of the kinds of remote interaction envisioned by Stelarc. These extended senses wouldn't be operated by pressing buttons or clicking on icons, though, but simply by thinking about them. Nicolelis argues:

> The brain knows that it has an arm and a hand because it is connected to these things and gets feedback from them. The same could be true for robotic or virtual appendages. If you control a remote hand that senses objects and sends tactile sensations back to your brain, it behaves as if it's your own hand. It becomes part of you. Your body becomes extended beyond the surface of your skin. Hybrid brain–machine interfaces have the potential to enhance our perceptual, motor and cognitive capabilities by revolutionising the way we use computers and interact with remote environments.

What Nicolelis is describing is a kind of reverse phantom limb syndrome. A phantom limb is defined as 'an arm or leg that lingers indefinitely in the minds of patients long after it has been lost in an accident or removed by a surgeon'. Many amputees continue to experience sensations or excruciating pain in their phantom limbs. Most people believe that phantom limbs occur because of spontaneous activity in the somatosensory cortex in the brain, particularly in the areas that represent the limb that has been lost.

Nicolelis speculates that new appendages, either robotic or virtual, could be added to the body and that the brain would eventually come to regard these as its own. In other words, the prosthetic limb would become a sensory add-on rather than an indication that something was missing, just as Stelarc suggests. 'The brain reflects internally the state of the body externally,' Nicolelis explains. 'Studies of brain plasticity [the brain's ability to reorganise itself] show that neural maps of the body are dynamic, continuously changing and adapting. If we created a brain–machine interface, a virtual or robotic object could become part of the brain's body map. We could augment our senses in virtual space in ways we never thought possible.'

Something very similar to these extra appendages was described by Samuel Butler in *Erewhon*. In Erewhon, machines were to be 'regarded as a part of man's own physical nature, being really nothing but extra-corporeal limbs . . . The lower animals keep all their limbs at home in their own bodies, but many of man's are loose, and lie about detached, now here and now there, in various parts of the world.'

What could a person do with a remote robotic or virtual limb lying about here and there in various parts of the world? The possibilities range from the mundane to the other-worldly.

In the virtual realm, Web shoppers could not only sniff the scent of a peach online, as we saw in the chapter on 'Smell', but squeeze it to see if it's ripe. Video conferences and chats might start with actual handshakes, and friends could greet one another with an affectionate peck on the cheek or a hug. And of course, there's always sex. Consenting adults could use the technology to engage in far more intimate embraces and manipulations. In the realm of robotics, devices could be sent to dangerous or inhospitable climes, such as deep-sea hydrothermal vents or the craters of active volcanoes, where researchers could gather specimens. Or imagine landing a sturdy robot on Mars; a scientist could amble around the surface of the planet picking up rocks and looking for signs of life, all from the comfort of mission control.

If Nicolelis is correct that virtual and robotic appendages can become part of the brain's body map, then in the future bodies may no longer be limited to two arms, two legs, two eyes and two ears.

We could have three arms, six legs and dozens of remote ears and eyes. Robotic appendages located in remote locations would be as close as our fingertips, as would virtual limbs situated in cyberspace. Someone in Tokyo could literally reach out and touch someone in Topeka over the Internet. And with enough practice, people would be able to generate the brain signals that controlled these extra limbs without moving a muscle – just by thinking about it.

Meet H.P.S., a 'locked-in' patient suffering from amyotrophic lateral sclerosis, a neurological disease that causes complete paralysis while leaving the mind intact. He is one of the first to explore the possibilities of mixing body and machine in this way. The disease struck H.P.S. in 1989 and his paralysis continued progressively. Since 1994 he has been artificially fed and ventilated. Owing to the disease H.P.S. has been unable to move or communicate for the past seven years; but thanks to the Thought Translation Device (TTD), a machine developed by Niels Birbaumer, a medical psychologist at the University of Tübingen in Germany, he can now dictate letters to a PC. 'Dear Mr Birbaumer,' H.P.S. wrote in his first full message composed with the TTD, 'I thank you and your team ... because you made me an ABC learner who often hits the correct letters.'

H.P.S. was able to 'hit the correct letters' because he learned to produce voluntary changes in his slow cortical potentials, the electrical signals in the brain that precede cognition or action. Two electrodes, placed on his skull above the area of the brain that controls motion, were connected to a computer that displayed the alphabet and a flashing cursor. The electrodes monitored changes in H.P.S.'s slow cortical potential rates linked to specific cursor movements: one rate caused the cursor to go up, another to go down. Once H.P.S. learned to control these signals at will, he was able to select letters of the alphabet and compose sentences by moving the cursor just by thinking.

Birbaumer and his team are adapting the technology, and speeding up the composition process by adding software that completes words after only a few letters have been selected, to allow patients to control everything from light switches to medical equipment to household appliances. Some patients could open doors or switch on the television with the TTD. One 'locked-in' patient in the US can perform actions similar to those of H.P.S., but the US patient has electrodes implanted in his brain.

Researchers at DERA, the British government's Defence Evaluation and Research Agency in Farnborough, are using a similar technique to help fighter pilots fly jets by thought power. DERA's 'cognitive cockpit' research programme is devoted to giving pilots the ability to control key flight systems simply by looking at or thinking about the appropriate icons on a computer screen. As with H.P.S. and the TTD, the pilots will first have to produce specific brain wave patterns on demand – the mental equivalent of tuning in to a radio station – to operate the plane's navigational or weapons systems. This technology would not only permit hands-free operation of crucial functions, but also give pilots quicker response times during complex and often dangerous manoeuvres.

Pilots inside a cognitive cockpit will have electrodes embedded in their helmets to monitor brain waves and an array of bio-sensors woven into their clothing to monitor vital signs such as heart rate and blood pressure. These data, along with information from the plane's other electronic systems, are fed into the cockpit's artificial intelligence software, which is designed to take on the role of a co-pilot. Since the cockpit computer will know the physical and mental state of the pilot far better than a human co-pilot could, it will be able to analyse his or her decisions and offer alternative solutions or draw attention to unexpected consequences.

When under attack, for example, the cognitive cockpit could alert the pilot to the urgent need for a navigational adjustment. Without being distracted from battle, the pilot could respond by quickly glancing at the relevant icon. Or should a pilot temporarily black out at high altitude, the cockpit would know this via data from its bio-sensors and could take control of the jet until the pilot regained consciousness. Pilots inside cognitive cockpits need to keep a close check on their thoughts, however, since a casual daydream or sudden recollection could have unintended consequences, such as inadvertently firing a missile.

Technologies like DERA's cognitive cockpit and Birbaumer's thought-translation device are still in the early stages of development and are a far cry from the virtual limbs envisioned by Nicolelis. For the human brain to truly incorporate prosthetics into its body map will require more than just plain thought: it will require feedback. The brain will only become aware of its new appendages if those appendages make their presence known.

To see how the monkeys might respond to this kind of anatomical extension, Nicolelis is creating a feedback loop between the monkeys and the robotic arm. In the next series of experiments the monkeys will have piezoelectric sensors attached to their bodies, so that the robotic arm delivers tactile sensations directly to their skin. When the monkey's brain waves impel the robotic arm to grasp a piece of fruit, for example, the animal will be able to feel the fruit's texture. The monkeys will also be able to watch the robotic arm in action on a computer screen. With this kind of visual and tactile feedback, Nicolelis hopes the monkeys will learn to associate the movements of the robotic arm with the achievement of their own goals. Once they make the link between the thought, the action and the result, they might not even take the trouble to stretch out their arms any more. Why bother when a mere thought will instruct the robot arm to do it for you?

Nicolelis admits, though, that there is still no non-invasive technique for tapping into these brain signals in humans, and without such a technique, these types of experiment are unlikely to be approved in people. The implants in the monkeys, for example, require extensive neurosurgery, not a procedure many people are likely to wish to undergo, even if the result could be the ability to operate computers by thought alone.

There is another potential problem as well. Nicolelis has demonstrated that it's possible to use brain waves to operate robotic devices; but what would happen if this process was reversed? The same signals that are currently routed from the monkey's brain through the computer to control the robotic arm could be sent back to the monkey to control its behaviour. Just as Stelarc became a puppet whose strings were pulled by computer users, monkeys with such implants could be manipulated according to the signals that we transmit to their brains. Such implants in human beings would likely face strong opposition unless the possibility of this kind of mind control could be eliminated, something that would be practically impossible to achieve.

Nicolelis is confident that a technological breakthrough will come, perhaps in the form of some kind of permanent intracranial implants, and that the ethical issues surrounding the technology will be resolved; but it is still likely to be some time before this merger between ourselves and our computers will be realised. When that day does come, however, it won't be the keyboard or the mouse but the thought that counts.

NEW ANATOMIES

Hugh Herr is part of a team at MIT trying to do for the human leg what Sinkjaer, Andersen and Nicolelis are trying to do for the human hand and arm. Herr and his colleague Gill Pratt run the MIT Leg Lab, which builds walking robots as well as artificial knees and ankles for amputees.

Unlike Sinkjaer's prosthetics, though, the MIT Leg Lab's devices are external, driven by microprocessors and outfitted with sensors to detect factors such as speed, force and torque. The artificial knee for amputees, for instance, acts as an interface between the residual limb and the artificial leg below it. Using information from the sensor array, the knee automatically adjusts its movements to create a smooth biological gait. It also adapts to new circumstances: if the user puts on different shoes or lifts a heavy suitcase, the knee responds accordingly.

Herr brings more than just technical expertise to this work. In 1982, at the age of seventeen, he lost both legs below the knee after a rock-climbing fall left him stranded for four days on Mount Washington in New Hampshire. By the time he was rescued, frostbite had so severely affected his legs that they had to be amputated.

Herr describes the accident as a 'life-defining experience'. Before the fall he had been a poor student, but an avid mountain-climber and athlete. Dissatisfaction with his artificial limbs made him determined to build a better prosthetic himself. So he went back to school to study computer science and physics 'to improve my condition and the conditions of others', he says.

To this end, Herr, Pratt and the Leg Lab's students are building bipedal robots that ambulate and prosthetic devices to help people walk. The goal is to replicate as closely as possible the abilities of human legs and then to incorporate these abilities into artificial limbs.

Herr is also involved in another robot project that doesn't have any legs at all: it's a fish.

The robotic fish — which looks like 'a five-centimetre surfboard with a tail,' according to Robert Dennis, a biomedical engineer at the University of Michigan who is working with Herr on the project — swims around like a bath toy on the surface of a tank in the biomechatronics group wet lab nine floors above the Leg Lab. Herr and Dennis started the biomechatronics group to study the basic science and engineering involved in the creation of hybrid biomimetic devices, such as the robotic fish, and ultimately hybrid prosthetic devices.

The fish is controlled by a microprocessor and operated by an infrared remote control. Press one button and the fish goes fast, another and it goes slow; press a different button and it turns left, another and it turns right. The amazing thing about this robot — and what makes it important to the development of better prosthetic limbs — is the fact that the tail paddles thanks to living muscles and tendons taken from a frog.

Muscles are wonderful motors — quiet, light, versatile, robust and efficient. What better way to mimic the agility, strength and speed of biological legs than to incorporate some biological muscles into the prosthetic? The biomechatronics group, with the Artificial Intelligence Lab at MIT, is building just such biomimetic devices, small robots like the swimming surfboard that are powered by living muscle tissue derived from animals.

In the robotic fish, the amphibian muscle and tendon tissue is sutured to a hard plastic spine with an elastic tail. Four cuff electrodes are wrapped around the neuromuscular juncture where the nerves and muscle meet. Power to operate the microprocessor and the muscle-stimulator circuitry is derived from a small battery the fish wears on its belly. In another twenty-first-century twist to Luigi Galvani's experiments, the frog muscles contract to move the tail when electrical stimulation is delivered via the remote control. The result: a hybrid frog-robot — part living tissue, part lifeless metal — that swims.

One of the most immediate problems related to the care and maintenance of meat–machine hybrids is this: what do you feed a robotic fish? Like all living things, muscle tissue needs nutrients to survive. Clearly, sprinkling a few dried goldfish flakes into the tank wasn't going to do the trick. So the biomechatronics group simply threw a few grams of glucose into the water so that the muscle tissue could extract energy from it. The batteries provide the electrical stimulation and the fish swims through its own food.

Over time, Dennis and Herr want to design hybrid prosthetics with an increasing content of living tissue and fewer synthetic components. The goal is to be able to grow muscle tissue on a Petri dish and use it in a machine. 'To finally realise this vision, many new technologies will need to be developed in parallel,' says Dennis, 'control of muscle phenotype in a bioreactor, nerve–muscle synaptogenesis and control *in vitro*, and engineered muscle–tendon interfaces, to name a few. We're currently working on it.'

Home-grown limbs, which could be cultivated from living human cells through a process known as tissue engineering, are still a distant prospect. In the meantime Gerald Loeb of the University of Southern California, Los Angeles, who did pioneering work on the cochlear implants discussed in the chapter on 'Hearing', has devised bionic neurons (or BIONs™) that might one day help researchers like Herr and Sinkjaer create a better nerve–machine interface. Loeb currently uses BIONs for therapeutic electrical stimulation to prevent and in some cases reverse the muscle atrophy that occurs in stroke and arthritis patients.

When a stroke victim loses the use of an arm, for example, the muscles in the arm fall into desuetude and gradually begin to waste away. Whereas functional electrical stimulation is designed to restore movement to such paralysed limbs, therapeutic electrical stimulation is meant to keep the muscles in shape by using electrical stimulation to activate nerve impulses to them. BIONs – self-contained, hermetically sealed stimulators about the size of a grain of rice that are injected directly into a patient's muscles – administer this miniature physical regimen by regularly sending mild jolts of electricity into the affected muscles.

BIONs are powered by a transmission coil worn in a sling on the patient's body and controlled by a Personal Trainer™. The Personal Trainer is a hand-held computer that contains three different exercise programs that vary the strength, timing and duration of the BION stimulation. The strenuousness of the workout depends on what kind

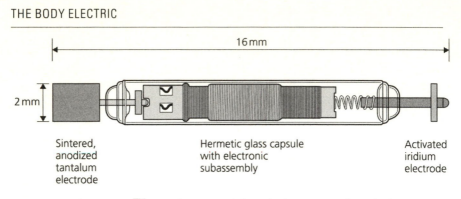

Figure 11. The BION™ stimulation module, which consists of a cylindrical glass capsule with two electrodes mounted on the ends.

of shape the muscles are in. The patient takes the Personal Trainer home, starts the programme, sits back and relaxes while the BIONs put the relevant muscles through their paces. According to one of the four subjects who have so far been injected with BIONs, the sensations thus created range from mildly pleasant to easily ignored.

Loeb is currently working on second-generation BIONs that will be capable of sensing things such as muscle length and limb acceleration. By transmitting this information back to the control unit, the BIONs could be used for functional electrical stimulation as well. Ultimately, Loeb and others would like to create BIONs that power themselves from an internal rechargeable battery and autonomously generate their own training regimens based on sensory feedback.

While the technology behind these neural prosthetics is impressive, it's important to point out that these devices are not cures for paralysis. There is an enormous difference between Brian Holgersen's ability to hold a coffee cup, for example, and the complete restoration of his sensory and motor abilities. The Freehand he uses restores only limited function, and even then only because he retains some movement in his hand and shoulders. The Freehand system, Sinkjaer's sensing system, Nicolelis's brain implants, Dennis and Herr's hybrid limbs and Loeb's BIONs all promise to greatly improve patients' quality of life, but they fall far short of anything that even remotely resembles a cure.

There are also complex technical problems to be solved before more sophisticated nerve–chip interfaces can be devised. Ironically, one of the main challenges is to make prosthetics that can withstand the hostile environment inside the human body. 'The most likely sources of failure tend to be in the physical connections between components,' says Loeb. 'This is particularly so when these connections must function in a pool of warm salt water that is constantly moving.'

It will be many, many decades before more robust and sophisticated prosthetics emerge, devices that would allow us to speak of an eventual cure for paralysis. Though, as Loeb makes clear, 'cure' is perhaps the wrong word. 'I don't think any of these technologies can be said to aim to "cure" paralysis any more than a cochlear implant "cures" deafness,' he says. 'Neural prostheses aim to restore function, not to cure the underlying cause of the dysfunction.'

'I dream of the day when there is no such thing as physical disability,' says Herr, who still works out once a week at a rock-climbing gym. 'Right now it's the technology that's disabled – it doesn't work.' Until this technological disability is overcome, people like Brian Holgersen will have to be content with the fact that greater independence and quality of life, if not a cure, are finally within their reach.

FEELING MACHINES

Electronic implants and neural prosthetics are not the only ways in which people are getting in touch with technology. Since emotions are crucial to understanding human beings, computers need to sense our moods, likes and dislikes, and idiosyncrasies if they are to serve us well. So Rosalind Picard and her colleagues in the Affective Computing group at MIT's Media Lab are giving them a crash course in emotional intelligence.

Affective computing involves endowing computers with the ability to recognise, respond to and, in some cases, even display human emotions. 'As emotions are important for human–human communication,' says Picard, 'they are important for human–computer communication. A quantum leap will occur when computers are able to recognise and express emotions.'

Sadly, the most common emotions currently aroused by computers are frustration, anger and despair. Consider the unfortunate user having problems with new software: her installation disks are buggy, her computer crashes, her temper flares. How much better if the computer could sense the user's rising anger – and do something about it – before it reached breaking point.

Picard and her team are developing new software that aims to do just that, reacting to the user's emotional state with a judicious mix of empathy, contrition and, when appropriate, helpful advice. The computer will, for example, solicit information about the user's frustration in the form of a simple pop-up window when a breakdown occurs. If the user expresses interest, the computer will try to suggest an alternate course of action in a way that is carefully scripted to avoid the pretence that the computer knows best. If there has been a particularly egregious fault, the computer will even apologise. 'Computers can do a good job of frustrating people,' Picard says. 'But if computers have more emotional intelligence they can also actively help people reduce their frustration level.'

How can a computer tell if someone is happy or sad, frustrated or fascinated, angry or absorbed? The same way as people do: by using their senses to look, listen, feel and learn. Affective computers are outfitted with electronic senses: video cameras watch gestures and facial expressions, speech-recognition devices monitor voice intonation, and a network of biosensors – unobtrusive and lightweight computers that are embedded in everything from clothing to jewellery – keep track of physiological signals such as pulse, respiration and skin conductivity.

Physiological changes provide clues to a person's emotional state. Dilated pupils, for example, may indicate arousal and interest, as may a quickening pulse rate. A low tone of voice suggests fatigue or depression, while slumped shoulders mean dejection. The computer, which through its network of biosensors is in constant touch with the user, picks up this information, and as a result is able to infer cues that suggest how a person is feeling. Instead of constantly asking the user to feed it information – by selecting preferences, scrolling through menus and clicking options – affective computers get to know their users by observing and responding to emotions. Of course, if the user prefers not to give the computer access to this information, the device can solicit it by asking.

Picard and her team have devised a number of innovative applications for this technology. Though not all of these applications directly involve the sense of touch, they are grouped together here because they do put users and computers in a more intimate union – so close, in fact, that in the near future we may never be out of touch with our computers. Wherever we may be, whatever we may be doing, they will be near, watching, listening, taking the pulse of our emotions and moods.

One such application could give a new twist to the term 'easy listening'. An affective stereo system, involving a 'wearable DJ' that monitors skin conductivity, could offer music based not only on the user's stated preferences, but also on the computer's interpretation of the user's mood as derived from its analysis of the relevant physiological signals. Or, alerted by a sudden jump in your skin conductivity, a biosensor embedded in a computer mouse could detect when a specific website grabs your interest.

When aroused by something, humans tend to perspire, and perspiration conducts electricity. So when you came across an exciting website, the computer could detect the increased skin conductivity in your hand and suggest similar sites that might interest you. The same thing could happen while you were strolling through an art exhibition. If you were captivated by a particular painting, the computer would notice and set off on a Web search for other exhibitions by the same artist. 'Affective computers try to perceive what you like and dislike and make suggestions on that basis,' says Picard. 'The computer takes the initiative to communicate with you. It tunes itself in to your preferences instead of expecting you to program it.'

At the Media Lab, Picard and company have come up with working prototypes that begin to tune in to some aspects of human emotional broadcasting. Not all of them are practical – indeed, some may be more trouble than they're worth – but they do suggest the range of possible applications for affective technology. The affective earring, for example, looks like an ordinary piece of jewellery, except that it contains a tiny non-invasive blood pressure sensor. When your blood pressure goes up, a potential sign of stress or excitement, the sensor relays this information to the computer. The Media Lab group has also developed a skin conductivity sensor that fits in a ring or a pair of shoes. The more discreetly such devices can be secreted around the body, the more comfortable people will become with their presence; and comfortable people are much more likely to give off relaxed signals.

These devices could have valuable medical applications as well. Media Lab researchers are building a family of wearables that measure, store, analyse and transmit physiological data such as heart rate, blood pressure and body temperature. This kind of information could be collected and monitored – either by the patient himself or remotely via the Internet by a healthcare professional – for early warning signs of illnesses such as heart attack or stroke. ElectroTextiles, a British firm, already makes a range of wearable fabrics that conduct electricity. The company is weaving this conductive cloth into a range of clothing in which the electronics, whether a cell phone, a CD player or a heart rate monitor, come as built-in fashion accessories.

Of course, while a heart patient might welcome the convenience of receiving blood pressure updates from his shirt, some people might get hot under the collar thinking that an insurance company or employer might get access to that same data too; and while having a stereo system woven into the lining of your leather jacket may be cool, consumers may be less comfortable with the fact that marketeers are monitoring the music they download from the Internet.

Affective technology enables users to give computers unparalleled access to their bodies, and the information thus gleaned may be transmitted over and stored on a network. For these devices to work, some of our most intimate details, such as heart rate, perspiration levels, muscle tension – physical signs of our emotional states – must be instantly available. What if someone hacks into the network and eavesdrops on your moods and feelings? What if medical information is obtained in this way and used against you during an application for a job or an insurance policy? Even now many people are reluctant to shop on the Web for fear that credit card information will be stolen. Before affective computers ever gain widespread acceptance, security and privacy will have to be assured.

One affective application in which the people involved do want others to know what they're feeling follows the theory that while the eyes may be the windows to the soul, the eyebrows are the peepholes to the emotions. Raised eyebrows can indicate astonishment or delight, while a furrowed brow suggests anger or confusion. The Media Lab's expression glasses are designed to give the computer a clear view of what a user's eyebrows are up to so that it can adapt its behaviour accordingly.

To catch emotions at play, the frames of the glasses include a small sensor that monitors muscle tension and movement in the eyebrows and forehead. After a couple of minutes of training, the glasses are capable of learning an individual's unique patterns of expression, accurately discriminating between feelings as different as surprise and bewilderment.

Why would anyone want to make a spectacle of themselves by actually wearing expression glasses? Because letting a computer – or another person who may not be in the same room – know whether you are confused or captivated could be useful. Imagine taking part in an important business meeting by video conference. Since it's often difficult to pick up subtle changes in facial expression by video alone, extra information on whether your counterpart is baffled or enthusiastic would be welcome. The expression glasses could do that by translating its observations into a visual display: raised eyebrows might be displayed as green to indicate interest, while a frowning brow might be shown red to warn of anger.

A teacher lecturing to a group of students by video link could find expression glasses useful for the same reason. The affective specs could serve as a barometer of comprehension levels among members of the class. If the glasses detected dozens of furrowed brows, the teacher would be able to adapt the lesson – by spending more or less time on certain subjects, for example – based on this feedback.

The Media Lab's 'startlecam' is devoted to more recreational pursuits. The startlecam is a wearable personal video camera, continuously in record mode, connected to a computer and a network of biosensors that monitor skin conductivity. It's designed to circumvent that bane of all new parents or holidaymakers: the need to grab and activate the camera to catch baby's first steps or that glorious tropical sunset.

The startlecam will record those precious moments for you – continuously if you want, since it can be set to 'always-on' mode – but since continuous recording is likely to collect a lot of boring moments as well, you can set the device to automatically mark the data it records with an indicator of your excitement level. When it detects a surge of excitement and matches it with other telltale signs of arousal, a buffer of images recently captured by the video camera – embedded in a shirt button perhaps, or on the front of a hat – is automatically saved. The result: instant home videos without ever having to worry about forgetting the camera or leaving the lens cap on. The images can even be downloaded onto your website or sent to the folks back home via e-mail as a digital video postcard.

Other researchers at the Media Lab are working to make entire rooms sensitive to what goes on inside them. These 'smart rooms' are strewn with cameras, microphones and other sensors – the walls really do have ears – capable of recognising who is in the room and what they are doing. Such an intelligent office could become a kind of room-sized personal assistant that knows your work habits, remembers where you put things, and is able to make phone calls, arrange meetings, book flights or retrieve documents acting on an understanding of your voice commands and body language.

Or the intelligent office could become a kind of room-sized surveillance system in which your every move, word, facial expression and even heartbeat is monitored, recorded and stored in a databank. Yes, you'll be able to ask the room: 'Where did I leave my notes from the meeting,' and it will reply: 'Under the pile of newspapers on your desk.' But is that such an improvement over simply rooting around until you find the notes yourself, especially given the fact that so much else of what goes on in your office, from private phone calls to embarrassing personal habits, will be available for scrutiny as well?

Proponents of technology agree that safeguards for privacy are needed. They also argue that you can simply turn the room off when you don't want it to record; but having to consciously decide to turn these types of devices off defeats the purpose of their ubiquity in the first place. The convenience of affective computing lies in the fact that users don't have to tend to these devices because they are always on and always responding to our signals. If users want them turned off, shouldn't this happen automatically too?

What if you forget to turn them off? Or what if someone else turns them back on without your knowledge or consent? How would you know? And what if the privacy safeguards are breached? 'Having computers observe our every move sounds like a nightmare to me,' says Picard. 'Most of us wouldn't want people around this much, so why would we want computers around us this much? We have a long, long way to go before we suffer such ongoing contact with any entity, no matter how smart. This is why wearables [as opposed to computers embedded in the environment] seem more empowering. We can easily turn them off or take them off. Our control over them is clear.'

Startlecams and expression glasses may be amusing and perhaps even useful, but it's still unclear whether anyone other than computer researchers will want to use them. Despite their efficiency and convenience, many people might react to the possibility of observant and emotionally intelligent computers with consternation and alarm. 'The truly emotionally intelligent machine will show respect for people's fears and concerns,' says Picard. 'If the user is dissatisfied with the machine's behaviour, the machine should offer to help the user to find a replacement for it, or offer to turn itself off.'

LET'S GET PHYSICAL

While Picard works to help computers understand what's going on inside human beings, Hiroshi Ishii is working to put people in touch with what's going on inside computers. Ishii, an associate professor at the Media Lab, is developing haptic technology that allows users not only to see and hear what's inside a computer, but to feel it, too. His 'tangible user interfaces' give shape to intangible digital information so that it can be directly manipulated and felt. The ability to generate tactile sensations has already been added to devices such as computer mice, joysticks and the dashboard knobs in automobiles. Ishii is trying to go beyond these devices, which for the most confine sensation to computers or other bits of technology, to give physical form to information itself. As haptic appliances spread, compelling websites might really feel sticky – and pixels could tickle.

Ishii originally got in touch with the physical aspects of information when he discovered his first personal digital assistant at the age of two: an abacus. He liked its smooth texture, the clear 'clink' sound it made when the beads were moved. 'As a young child I remember the joy of using an abacus as a toy train, a musical instrument, even a backscratcher,' he says, 'or of being aware of my mother's presence merely by hearing the ambient sounds of the abacus beads clicking as she figured the household accounts.'

Inspired by that childhood experience, Ishii has been devising ways to make the wealth of digital information inside computers as tangible and pleasant to the touch as the abacus of his youth. 'Now we experience the digital world through typing and clicking on cheap plastic boxes,' he laments. 'People care about touch and appreciate things in the real world – a nice shirt, a fine fountain pen. But the digital world is separate, trapped inside the computer screen. I want to use objects that we can touch to manipulate digital information, to join the richness of the physical world with digital technology. The beauty of the abacus is that digits (information) are represented physically so that users can manipulate them directly.' The abacus is the key model for what Ishii is trying to achieve, because it makes no distinction between input and output: the beads, rods and frame are both physical representations of numerical information as well as computational and control mechanisms.

Ishii's lab is cluttered with stuff: board games, Lego bricks, badminton rackets, and a clutch of unidentified projects in various states of dishabille. In fact, the place looks a lot like my six-year-old son's bedroom after an afternoon of serious fun – which makes sense since Ishii subscribes to the school of learning-by-playing. 'It's easier to understand something if you're able to play with building blocks like a kid,' he says. 'To actually physically feel something enriches our perception of it and allows us to interact with the environment as we were meant to.'

Ishii's Tangible Media Group is developing haptic interfaces: systems that allow users to reach out and touch someone or something through a computer. These tangible user interfaces employ physical objects, surfaces and spaces to embody digital information. 'We live between two realms,' Ishii says, 'our physical environment and cyberspace. The absence of seamless couplings between these parallel existences leaves a great divide between the worlds of bits and atoms.'

The inTouch system, a device that creates a direct link between two people separated by physical distance, is one device that bridges this great divide. Using 'force feedback' technology, in which physical pressure is exerted through computer networks, the inTouch system brings people together through the manipulation of two identical but remotely located objects.

The current inTouch prototype consists of twin sets of three cylindrical wooden rollers connected to a computer. The rollers are haptically coupled so that each one feels as if it is physically connected to its counterpart. Sensors monitor the positions of the rollers and precision motors synchronise their movements. When a user rotates one set of rollers, the computer rotates the corresponding set in exactly the same manner. In this way someone operating the inTouch system actually feels the movements of her counterpart, even though she may be located in another city. By means of telemanipulation technology like this, these synchronised, distributed physical objects create the illusion that distant users are interacting with shared physical objects.

Users of inTouch can passively experience the other person's manipulation of the rollers, collaborate to spin the rollers in the same direction or struggle for sole control. With the addition of biosensors like those used by Rosalind Picard, it might eventually be possible to detect a user's emotional state according to the type of pressure exerted through the device. More speculative possibilities include a tangible telephone, through which callers could actually feel each other on the line. Phone sex would suddenly take on a whole new meaning.

Tangible telephones, kinaesthetic phone sex. It all feels good, but does it have any practical use? Ishii is convinced that architects, for one, would find tangible user interfaces extremely practical. Two architects collaborating on a project, one in New York and one in Tokyo, could actually manipulate the same scale models to experiment with different designs, see how pedestrian traffic might flow through streets, or how the height of buildings might cast shadows or create wind tunnels. They would be able to both move and see others' movements of the same objects.

Ishii's musical bottles are perhaps his most magical and potentially useful devices. Dozens of bottles line a wall in a dim corner of Ishii's lab: old brown medicinal ones, crazily coloured ornamental ones, elegant crystal ones, even ones Ishii has blown himself. Take a bottle from the shelf, place it on a sleek black table, which is illuminated from below with soft glowing purples and blues, and lift the stopper. Depending on which bottle you choose, the sound of birdsong in Sapporo, Ishii's home town, pours out; or perhaps you'll get the Cambridge, Massachusetts, weather forecast; or if you pick the right combination of three related bottles, you'll get a jazz trio consisting of piano, violin and cello.

The idea behind the bottles is that laptops and desktop PCs are not the only places in which digital information can, or should, reside. In Ishii's view it's more practical to put that information where people need it most: in the objects themselves.

Each of Ishii's singing bottles contains a sensor that, when placed on the table, imparts instructions to a PC. The PC follows those instructions and pulls up the appropriate information, whether it be the twittering of Sapporo's birds or a Thelonius Monk recording. Ishii suggests talking pill bottles as one useful application of the technology.

When a patient is taking medication, the bottle could remind him of the appropriate dosages or possible side effects every time he popped the lid. This could be especially helpful to elderly patients who may have trouble reading the label or keeping track of their medication regimen.

Ishii envisages more evocative uses as well. The next time you give a gift of perfume, for example, why not program the bottle to play a favourite song or recite a poem when it's opened? 'People have used glass bottles for thousands of years,' Ishii says. 'Why not use them to hold digital information? It's a way to think beyond the computer screen and keyboard to find more natural ways to access information.'

In 1993 Kenneth Salisbury and Thomas Massie, then at MIT's Artificial Intelligence Laboratory, were thinking beyond the screen and keyboard when they invented the Phantom haptic interface. The Phantom, the robotic arm used in Nicolelis's monkey experiments, consists of a stylus mounted on the end of a high-tech jointed appendage that lets users feel the information inside computers.

To operate the Phantom, just pick up the stylus and it reproduces your movements on a computer screen inside a virtual space filled with geometric shapes. When you touch something in that space – the tip of a triangle, for instance – the Phantom exerts force on the stylus so that you actually feel its sharpness. By modifying the force and adding irregularities, even rough and smooth textures can be reproduced. 'Using the Phantom is like feeling the world with a stick,' says Mandayam Srinivasan, head of the Laboratory for Human and Machine Haptics (or Touch Lab) at MIT and one of Nicolelis's collaborators; but it's a remarkable stick, nevertheless.

The Phantom looks like a cross between a postmodern Nordic desk lamp and a vaguely sinister dentist's drill. Using the Phantom haptic wand at Srinivasan's Touch Lab, I was able to roam through a virtual room filled with floating geometric shapes. Prodding a big, bulky rectangle I could feel its weight as I shoved it into a passing triangle. I felt the thrust and rebound of the collision as the shapes bounced into opposite corners of the screen. Rubbing the stylus across a ribbed oval shape produced the same bumpy staccato sensation as running a stick along a picket fence.

I also manipulated virtual models of the lungs and liver. I could see the tissue squish and dimple as I poked and prodded. When I tugged on the liver with a pair of virtual tweezers, I felt the resistance; when I let it go, it snapped back into place like a rubber band. Medical training is, in fact, one of the primary applications for haptic interfaces such as the Phantom. The idea is that young surgeons first perform virtual operations on computer-generated organs before moving on to the real thing.

Impressive as these sensations were, though, they were less than entirely convincing. The touch impressions were somewhat artificial and muted, like the feeling of a dentist's drill under novocaine: you feel the pressure of the drill as if it's very far away. Unlike actually touching a solid object, the sensation is more like that produced when you hold your hand out of the window of a speeding car and let the wind press it back. Using the Phantom, somehow I couldn't escape the feeling that there was 'no there there'. To increase the fidelity of the sensation, researchers at the Touch Lab add visual or auditory cues. For example, Srinivasan has experimented with adding sound effects, so that when you poke a rectangle you actually hear a tap or a thud or a clink.

Force feedback technology like that employed in the Phantom provides only the feelings you would experience if you touched the world through a rigid tool handle. To recreate the full sensations of touch tactile displays are needed. Virtual Technologies, Inc. in Palo Alto, California, already makes a tactile interface that allows users to reach in and physically interact with simulated objects inside computers. The CyberGlove is a lightweight glove embedded with eighteen flexible sensors that, when connected to a computer, produces authentic force feedback effects such as jolts, pulses and vibrations. The grasp forces are exerted through a network of tendons routed to the fingertips via an exoskeleton, known as CyberGrasp. Tactile

131

sensations are generated by means of tiny vibrotactile stimulators on each finger as well as the palm.

Putting on the CyberGrasp is a bit like affixing a spider crab to your fingers. When I tried one on, I first donned a Michael Jackson-like white glove to insulate my hand. Then I put on the CyberGrasp itself, which is a thick piece of cloth with the metal exoskeleton attached to each finger. I was then able to manipulate tools on a virtual worktable, which was displayed on a computer screen. When I picked up a virtual hammer, for instance, I could feel its weight. When I pounded it on a virtual engine component, I felt it thud and rebound in my hand.

Srinivasan compares a digital world without haptics like the CyberGrasp to 'life in a museum where you are not allowed to feel anything'. As information on the Web becomes more complex, he believes people will insist on having more natural ways of interacting with that information. If you buy a sweater on the Web, Srinivasan says, you'll want to touch it first. If you make a video call, you may want to touch as well as see your conversation partner.

'We push on the external world and it pushes back on us,' Srinivasan says. 'Yet in the world of computers our primary mode of receiving information is still visual and depends on abstractions like words and diagrams. Touch is too often taken for granted, but it's the only sense that is directly coupled to action in the world. We need haptics to get things done in the digital world, too.'

Michitaka Hirose and colleagues at the University of Tokyo are using haptics to merge the physical and virtual worlds. They have developed a portable haptic device, called HapticGEAR, that creates feelings of touch inside immersive virtual environments. Haptic-GEAR sports the same stylus as the Phantom, but all the technology is worn on a backpack instead of being tethered to a desktop. The force feedback sensations are generated by means of four tension wires emanating from the backpack. Rather than just viewing and interacting with a virtual environment on the screen, HapticGEAR allows users to roam around inside a computer-generated space and poke about with the stylus. A medical student, for example, can don HapticGEAR and stroll around inside a virtual liver while palpating tissue.

Hirose hopes to build even more elaborate and physically realistic worlds through the Multimedia Virtual Laboratory, a system by which two separate labs share the same virtual environment and the same visual, auditory and haptic sensations.

The two labs, one in Gifu near Nagoya and the other at the University of Tokyo, are connected via a broadband network by two highly sophisticated and very expensive versions of virtual reality environments called CABINs (Computer Augmented Booth for Image Navigation). These environments sport six screens – one each to the front, back, left, right, top and bottom – that completely surround the user by projected images. Three-dimensional virtual environments are created and projected into this space just as in other virtual reality systems. What makes the MVL different, though, is that a user in the Gifu lab can be present in and interact with the environment at the Tokyo lab thanks to telepresence technology.

This kind of teleportation is possible because each CABIN has about a dozen video cameras embedded around the room. The cameras capture the image of the user in the Gifu CABIN, while a computer extracts the user's figure from the background visual information. This image is transmitted in real time through a broadband connection to the CABIN in Tokyo, which displays a virtual environment identical to the one in Gifu. The video avatar of the Gifu user is then projected into the Tokyo CABIN. The telepresent person is no glorified video image either: it is three-dimensional and comes with all the relevant positional information, so users are rendered at their actual size and in proper relationship to the other virtual objects in the space and the positions of the other users. The user in Gifu appears as if he is physically present in the Tokyo environment.

As in Ishii's distributed architectural models, both users can see and manipulate the same virtual objects. What's more, they can see and interact with each other via their video avatars – the next best thing to being there. Using the Multimedia Virtual Laboratory, a medical student in Gifu could stroll through a virtual liver with an experienced surgeon in Tokyo inside the same computer-generated space. With HapticGEAR, they can touch and manipulate the same objects. 'Users have the sense of being in the same space and sharing the same world,' Hirose says.

Though systems like inTouch, Phantom and HapticGEAR are still a far cry from replicating, say, the subtle pressures and textures of a simple handshake, it's not hard to imagine the recreational uses to which this technology could be put once sufficient sophistication is achieved. The late Michael Dertouzos, formerly head of MIT's Laboratory for Computer Science, foresaw a new form of entertainment to complement audio and video: 'bodyo'. By donning virtual-reality bodysuits outfitted with an array of sensors and entering an environment like the Multimedia Virtual Laboratory, we will one day be able to feel everything from the softness of computer-generated grass to the pleasures of digital sex.

Bodyo may already be here. An inventor in New South Wales has designed lifelike robotic mannequins to act as surrogates for distant partners who want to engage in long-distance sex. The mannequins are high-tech versions of inflatable sex dolls. Covered with artificial skin, they are powered by servomotors and move in response to signals received from the Internet as well as from tactile and audio sensors distributed over their surface. To have sex, two people with complementary mannequins link up over the Net. The interactions of each partner with his or her own robot are transmitted to their partner's robot. For added realism, the lovers could wear virtual-reality visors so they can hear and see the object of their desire – as well as make him or her resemble the celebrity of their choice.

Haptics are also being put to other recreational uses. Through joysticks, steering wheels and game pads equipped with force feedback technology, computer game players can enjoy more authentic and sensational experiences. Accelerating to 290km per hour, virtual race-car drivers can feel the force of acceleration push them back into their seats. Virtual sharpshooters can feel their weapon recoil when they blow away some alien invader.

Will haptics research end up simply providing consumers with new gimmicks for computer games? Or will it eventually lead to richer, more sensual ways to interact with information and communicate with other people? Probably both. The human hand itself is incredibly versatile: hands can play the piano, dig a ditch, perform brain surgery, tie shoelaces, caress a cheek or break a nose. Why should haptic technology be any less diverse?

Haptics research will certainly extend the human hand's reach, across biological frontiers and into virtual worlds. Ultimately, though, 'the hand, like all the other senses, ends at the brain', observes the Touch Lab's Srinivasan. No matter where our sense impressions first originate – from the eyes, ears, nose, mouth or skin – their ultimate destination is the brain. It is only here that they are processed, perceived and endowed with meaning. Since the brain is where all our senses end it is where the next, and final, chapter begins: Mind.

SEVEN

MIND
The Sixth Sense

'This is an idea with which I have toyed before – that it is conceptually possible for a human being to be sent over a telegraph line.' NORBERT WIENER

In a garage somewhere in Somerset in south-west England a new genesis is taking shape. Artificial life researcher Steve Grand is building a baby orang-utan. But this is no ordinary orang-utan: Grand's robotic primate, called Lucy Matilda, will have the eyes, muscles, nervous system and brain of a biological orang-utan and he hopes to help Lucy work her way through nursery school, learning to coordinate her muscles, form spoken words and eventually paint pictures. 'Lucy will not be as smart as a human or ape baby of the same age,' Grand says, 'but she will learn for herself and she will have something that no robot has ever had before: an imagination.'

Lucy's body will be made of electronic sensors and motors, and her mind will be a neural network that mimics the functions of the brain. 'We are ready to return to the Garden of Eden,' Grand says. 'But this time we will not be mere produce of the garden, but gardeners ourselves. We are on the verge of being able to create life forms of our own.'

His claims may not be as outlandish as they seem. Grand intends Lucy to demonstrate that the term 'artificial life' is no oxymoron: all life, whether it's made of DNA or computer code, is real life.

In the Grand scheme of things, life is a garland of cause and effect, a recurring set of patterns, the most salient characteristic of which is simple: it persists. He would adapt Descartes's fraught proof of his own existence – *cogito, ergo sum* – as, 'I persist, therefore I am.' 'Patterns that persist by metabolising and reproducing are alive,' Grand says.

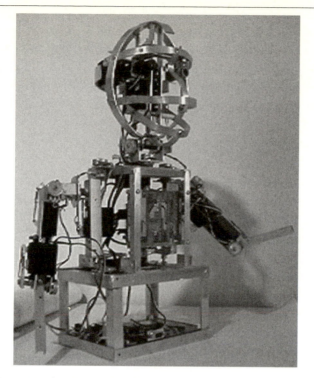

Figure 12. Lucy Matilda, Steve Grand's robotic primate.

Grand's work, and that of other researchers in the fields of neuro-science, artificial life and artificial intelligence, shows that Descartes's reasoning was flawed. Thought cannot, in fact, be abstracted from the body because our senses are the only avenues through which we perceive and interact with the world. Far from being superfluous distractions, what we see, hear, smell, taste and touch actually determines what and how we think. Consider, for example, how something as simple as an empty stomach can derail even the most logical train of thought. Without a body – including that miraculous engine of cogitation itself, the brain – there can be no awareness, intelligence or consciousness. In short, if you don't have a body you'll never have a mind.

The question of just what a mind is – or, in the opinion of some philosophers, *if* a mind is – lies far beyond the scope of this book. For our purposes, it is enough to define mind as comprising three qualities: awareness, of the senses and the information they provide; intelligence, to use this information to survive and thrive in the environment; and consciousness, the ability of an entity to reflect on its own awareness and intelligence and realise that it is separate from the world 'out there'. This chapter is about attempts to evoke these three qualities in artefacts constructed by people like Steve Grand, and about possible ways to link our biology-based mind with an artificial brain.

Since bodies are essential to the emergence of mind in human beings, it makes sense to assume that artificial creatures need bodies too if they are ever to become aware, intelligent and, perhaps some day, even conscious. The rallying cry for this kind of research might be summarised in the slogan, 'No sensation without representation.' Computer scientists are providing the necessary physical representations for these possible minds by taking computer pioneer Alan Turing's advice: give machines – both virtual ones inside computers and physical ones in the form of robots – the best sense organs that money can buy.

Gerald Edelman, winner of the 1972 Nobel Prize for Physiology or Medicine and a leading researcher into the neurobiological basis of consciousness, proposes a simple four-point plan to explore the mind–body question as it applies to artificial life forms:

'1. Simulate the organ or the animal – making provision for the fact that it contains a generator of diversity – mutations, alterations in neural wiring, or synaptic changes that are unpredictable.

2. Independently simulate a world or environment constrained by known physical principles, but allow for the occurrence of unpredictable events.

3. Let the simulated organ or animal interact with the simulated world or the real world.

4. See what happens.'

What happens, at least in Grand's application of these instructions, is Albia, the lush, jungle habitat of 'Creatures', one of the most popular computer games around.

Before he started building Lucy, Grand created Norns, the endearing, bug-eyed software agents that live, eat, mate, play and pass away in the Creatures games that run on about 1 million computer screens worldwide. Norns, first released into the wild in 1996, are sophisticated software programs modelled on biological processes. They have computer-coded genes and bloodstreams and live in Albia, where they interact and evolve and are subject to everything from dramatic mood swings to fatal diseases. Norns are patterns inside a computer that persist by metabolising and reproducing; and Grand, along with quite a few Creatures fans, believes they are alive.

The Norn metabolism replicates inside a computer what goes on inside a human body, and Grand's coding wizardry ensures that the Norn genome contains generators of unpredictable mutations and alterations. Norns are born with anywhere from 600 to 800 genes, bits of digital DNA that contain instructions that control the creatures' appearance, biochemistry and bodily functions. The genes comprised of this DNA mutate, become damaged, cause disease and are passed on to offspring, just as in biological creatures.

The Albian environment is as complex and diverse as the Norns themselves, and is constrained by its own physical laws. Albia itself resembles a tropical island, replete with an astounding variety of flora and fauna, including insect pests and something called the 'empathic vendor', a lotus-like vending machine that spews out a variety of edible geometric shapes.

Though it looks green and inviting, Albia is no paradise. It comes complete with its own socioeconomic problems. Norns must eat to survive, so they spend a lot of time foraging for food inside their virtual terrarium. Fortunately, a cornucopia of proteins, starches and sugars is provided by the local vegetation; but if the Norns eat too much too fast, famine results, and some Norns could die before the food supply is replenished.

If not of starvation, Norns can also die as a result of disease or getting the bits kicked out of them by a Grendel. Grendels are a race of hulking miscreants, violent sociopaths that disrupt the Albian idyll with all kinds of unpleasantness, from voraciously consuming all the available food to pummelling the life out of defenceless Norns. The Norns are not pacifists, however. When one particularly sadistic Grendel attacked a young Norn, the parents defended it. The father was killed in battle, but the mother survived and eventually slew the Grendel.

Assuming they escape famine, disease and the onslaughts of marauding Grendels, Norns interact with their environment through a full complement of senses. They can see, hear, smell, taste and touch. They can communicate with the user and with other Norns through text bubbles. The user types in a query, such as: 'How are you feeling?' and the Norn might reply: 'Really hungry for protein.' Norns also emit and can detect their own brand of pheromones – imperceptible odour molecules that can influence behaviour – as a way to inform members of the opposite sex that they're ready to mate.

Procreation is accomplished through osculation. Norns exchange genetic material during an innocent kiss that culminates in a loud 'pop' as their lips separate. A few minutes later the female lays an egg. A few minutes after that a little Norn is born.

I was fortunate enough to witness the birth of one little Norn during a visit to the offices of the company that produces the Creatures game, Creature Labs, in Cambridge, England. When it hatched, the baby Norn's first words as it poked its head out of the shell were 'Bub na daa' – like human infants, Norns must learn to talk.

Furthermore, at some point all Norns must die, of course. The average lifespan is only about ten hours, and the ageing process is regulated by something called the 'life chemical'. Each Norn is given a quota of the life chemical at birth, and as it slowly decays over time the Norns turn grey and wrinkly and finally expire. When they die, their bodies gradually decompose into a shower of shimmering sparks not unlike the transporter beam in the original *Star Trek* series.

Norns learn to walk and talk and generally get around in their world the same way as humans do: through a combination of nature and nurture. Nature, in the form of Grand's programming, provides the system of instincts and behaviours that motivate the creatures to act and learn. The creatures' neural and chemical structures are encoded in their digital DNA, which is passed on to offspring and is prone to mutation, so a kind of evolution takes place. As for nurture, a Norn is born with certain drives (to eat and reproduce, for example) and dispositions (some are active, some are lazy, some are sociable, some are stubborn); but how these drives and dispositions develop depends on the environment and the Norns' experience of it.

The technology behind Creatures is based on neural networks, the clusters of electronic circuits that work in ways analogous to the ways neurons work in the brain that were described in the chapter on 'Sight'. In a neural network, computer circuits replace the net of interconnected neurons inside the brain. Just as in biological brains, the Norns' electronic neurons strengthen or weaken their signals to and connections with each other in accordance with the stimuli received through the senses.

This neural traffic is regulated by chemo-emitters, which transmit the virtual neurochemicals that drive the Norns' behaviours and needs, and chemo-receptors, which receive these messages. The chemo-emitters and chemo-receptors function like biological synapses, which ferry instructions around the brain. When stimulated by sensory input, chemo-emitters fire off instructions that are picked up by the chemo-receptors and translated into specific actions or behaviours.

When a Norn picks up a savoury triangle from the empathic vendor and pops it in its mouth, for example, its neural circuits fire off messages to other circuits, essentially saying: 'This is tasty and nourishing. Remember, food can be found near the empathic vendor.' As this behaviour is repeated over time, it becomes reinforced and the Norn learns to associate food with empathic vendors. In this way a neural feedback loop is developed, and as a result the Norn strolls by the empathic vendor whenever it's hungry.

Norns are able to learn things about their environment – like the link between nourishment and empathic vendors – because their neural networks divide the world into two parts: what is it? and, what do I have to do about it? The answer to the first question is provided by what Grand calls 'concept neurons', which take sensory inputs and classify them into several broad categories. If a Norn eats something and notes as a result that its energy level rises, it files the object under the 'food' category. If, on the other hand, it eats something that does not boost its energy level or makes it sick, that object goes firmly into the 'non-food' category.

Once an object or a situation has been identified in this way, 'action neurons' kick in. If the aforementioned object did indeed boost the Norn's energy level, then the Norn will seek out more such objects. If not, the Norn won't bother trying to ingest one again. The Norn is not explicitly programmed to make specific associations to specific objects, however. It's programmed only with basic drives, such as 'find food' and 'stay alive'. The neural network does the rest. In short, the Norn learns from experience and so can make autonomous decisions ('Should I eat this?'), remember past encounters (with sociopathic Grendels, for example) and plan for different possible futures ('Next time a Grendel approaches, go to another part of Albia').

Grand believes that because the Norns can learn, remember the past and make autonomous decisions, their neural network 'deserves to be called a brain. It is pretty stupid in comparison to many animals. But it is a brain, rather than a computer program, and it does, in a rather limited sense, think. The thoughts are not programmed in – the behaviour of the whole structure is not immanent in any of its parts but an emergent consequence of many tiny, concurrent, "thoughtless" interactions.'

Grand is using such 'tiny, concurrent, "thoughtless" interactions' to make Lucy walk and talk; and since he's an engineer, not a scientist, he's trying to understand these new forms of life from the bottom up: by building them rather than by taking them apart. To move his life forms from the virtual world to the physical world, though, they will need two things: senses to tell them where they are and 'imagination', as Grand puts it, to tell them where they want to be. Grand is trying to endow his devices with imagination by 'putting the soul back into lifeless machines – not the souls of slaves, but willing spirits, who actually enjoy the tasks they are set.'

But where does imagination come from? And given that no one has the faintest idea how it works in the human brain, how could it possibly be inserted into a machine? Grand admits that he has only tantalising clues as to how the imagination works in human beings. What he does know, however, can be summed up by the phrase: the whole is greater than the sum of its parts.

'My mind exists because of the interaction of the billions of neurons in my brain,' Grand explains, 'yet no single neuron could be said to contain my mind. Only when the assembly is acting as a whole does my mind exist. Moreover, the existence of my mind would come as a complete surprise to anyone who was only given one of my neurons to study in isolation.'

As far as the brain is concerned, not only is the whole more than the sum of its parts, he stresses, but the whole cannot even be predicted by a study of the parts alone. Grand compares a neuron to a raindrop. It's not possible to imagine Niagara Falls from a single drop of rain, yet that immense torrent is just a bunch of raindrops acting in concert. 'Even though we feel in our heads like we are one person,' Grand says, 'we are really a loose confederacy of a hundred thousand million tiny, very stupid machines. Our brains are made of vast numbers of neurons, each operating in parallel. There is no central controller and no serial, stepwise process of execution. Intelligence is the result of thousands of millions of unintelligent processes operating concurrently.'

The production of complex behaviour from simple building blocks – Niagara Falls from a flood of raindrops, or imagination from a bunch of neurons – is known as 'emergence', and it is this principle that Grand is trying to replicate in Lucy. He hopes to achieve critical brain mass through the 'psychoprocessor', a computer chip with a mind of its own. The psychoprocessor is what Grand calls 'the basic building block of a mind': a single, general-purpose neural module, that when connected in suitable ways to thousands or millions of other psychoprocessors, learns to exhibit a wide range of behaviours, none of which are preprogrammed. By linking enough psychoprocessors together in the right ways, Grand hopes imagination will result. If it works, Lucy will learn to crawl and then walk, to babble and then talk.

'If you want a system that behaves like a small creature,' Grand reasons, 'then build a small creature by putting lots of dumb electronic neurons together to create an entity with a mind of its own.' The goal, as embodied in Lucy, is to create a general-purpose 'creature on a chip' that has a physical body and whose mind resides inside a computer. Baby Lucy can so far hear, see and make noises and she has moving arms and a head. As the robot grows up into Toddler Lucy, Grand hopes to add legs and more powerful muscles so the robot can learn 'many of the things that young babies have to learn: how to move one arm independently of the other limbs; how to reach out and grab; how to recognise simple spoken words and the mood of the speaker's voice; how to copy those sounds.'

So what? you might ask. What good would a robotic orang-utan be? It's easy to imagine a range of nifty applications for machines that can think: space exploration comes to mind and, closer to home, a host of uses in the industrial and military sectors. 'I want to build machines with minds of their own because these machines will be intelligent and adaptable, and can therefore be applied to many tasks,' Grand says. 'Instead of us having to adapt to our machines, our machines will adapt to us.'

If Grand succeeds in endowing his chips with imagination, though, he will have given computers that vital spark which, by current definitions, would qualify them as forms of life. If that happens, Lucy – named after the famous australopithecine skeleton that provided one of the missing links in human evolutionary history – may one day be placed at the root of another kind of bionic family tree.

Grand, however, is also careful not to overstate his claims. 'I do not believe these creatures are conscious,' he says. 'They are alive, but they do not have minds. If you knock on their door, I think you'll find there is nobody home.' Yet, he doesn't rule out the possibility of conscious machines, either: 'I believe that life can be created where there was none before. I think that it is possible to make thinking, caring, feeling beings and that, when these beings exist, it may be reasonable to ascribe to them a soul.' For Grand, life is – literally – what you make it.

IT'S WONDERFUL A-LIFE

Are Norns alive? If so, do they have minds and imagination? One way to begin answering such questions is to apply the famous duck test: if it walks like a duck and quacks like a duck, then it's a duck. If an artificial creature walks as if it has a mind and talks as if it has a mind, then it has a mind. Or does it?

The eighteenth-century French inventor Jacques de Vaucanson was one of the first to apply this method to artificial creatures. In 1739 De Vaucanson – who, among many other industrial innovations, automated the loom by means of a series of perforated cards, a technique later perfected by J.-M. Jacquard – created a robotic duck. De Vaucanson's automaton not only walked and quacked like a duck, but drank, ate, flapped its wings and defecated like one too. (De Vaucanson was the first person to find an industrial application for the rubber hose, in this case as part of the duck's digestive tract.)

Today, a new breed of automaton far exceeds the capabilities of De Vaucanson's duck. Sony's robotic dog Aibo can fetch a stick, perform amusing pet tricks and respond to simple voice commands. Other Sony robots can dance, jump and even play football, while Honda's Asimo climbs stairs, walks round corners and is pretty light on its feet on the dance floor as well. Few people would suggest, however, that any of these robots has a mind. Robokitty, however, may be different.

Robokitty is a life-sized robotic cat under construction by Hugo de Garis, who has been building his bionic feline over the past ten years at research labs in Japan and Europe. De Garis hopes that Robokitty may one day pass the duck test.

Though Asimo and Aibo may be mindless creatures, there is nothing to prevent mind arising from materials other than biological brains. Brains are necessary, but they don't have to be composed of organic tissue and biochemicals. 'Where consciousness is concerned, brains matter crucially,' philosopher John Searle writes in his book *The Mystery of Consciousness*. 'An "artificial brain" might cause consciousness though it is made of some substance totally different from neurons, but whatever substance we use to build an artificial brain, the resulting structure must share with brains the causal power to get us over the threshold of consciousness. It must be able to cause what brains cause.'

De Garis is trying to cause what brains cause by designing a silicon cerebrum built by a supercomputer.

De Garis fits the mad-scientist stereotype almost too well: baggy pants, ill-fitting sweater, amiable but distracted manner, and an unruly shock of hair that leaps from his scalp like a bunch of wayward dendrites. As he speaks, his right hand continually rakes his hair as if trying to dislodge an irate bee.

Also in line with the stereotype are De Garis's apocalyptic musings on the future. He says it's 'tragic to freeze evolution at the human level' and is sure that some time soon, within the next hundred years for sure, massively intelligent creatures will evolve from later versions of the types of brains he is making to take over the world, enslaving or wiping out the human race in the process. De Garis describes what he's doing as 'building gods'.

Alas, gods aren't built in a day, and De Garis has been at it since 1993, when he first proposed Robokitty, a mechanical kitten that he hopes will be able to walk, purr and do many of the things a biological kitten can do. Robokitty's brain will sport 75 million or so artificial neurons and will be housed in a supercomputer that will control the robot by radio signals.

If De Garis succeeds in building it, Robokitty will be different from bots like Asimo and Aibo because it will operate autonomously using its own artificial nervous system, in which each complex behaviour is executed by the robot's artificial neurons — and nothing else. Conventional bots are operated by software programs written by software programmers, who must specify exactly how the device is to perform each task. If the programmers want the robot to kick a ball, for example, they have to write a program detailing every phase of this behaviour: move right foreleg forward; shift body weight to left foreleg; raise paw of right foreleg; and so on.

The human brain, like the brains of other animals, is not a computer, however, and there is definitely nothing even remotely resembling a computer programmer — except, of course, the process of evolution. Whereas conventional computers operate in a rigidly logical and centralised fashion, brains are notoriously messy, decentralised and redundant and operate in a decidedly non-linear fashion. As we've seen in previous chapters, the brain is amazingly resilient; if part of it is injured or removed, another region will take over the damaged part's functions. If, on the other hand, a chunk of Aibo's programming was removed, the robot would immediately keel over.

De Garis wants to give Robokitty the brain of a cat, a brain that functions in the adaptive, non-linear manner of the biological model; but creating the complex circuits needed for such an artificial brain is beyond current human programming capabilities. Since human beings can't design the circuits themselves, De Garis is throwing the power of evolution at them in the form of the CAM Brain Machine (CBM).

Though its name evokes visions of some majestic robot from the classic silent film *Metropolis*, the actual CBM looked more like a disassembled air-conditioning unit when I saw one being put together. Though it may have an unprepossessing exterior, the CBM (one of only four in the world, each costing $500,000) grows and evolves intricate neural circuitry at incredibly high speeds. De Garis hopes to use the CBM to build Robokitty's brain.

To get his artificial feline up and running, De Garis reckons he needs a brain with roughly 64,000 circuits. Each circuit contains 1000 neurons, which amounts to a grand total of some 75 million neurons. (For comparison, the human brain contains about 100,000 million neurons.) These artificial neurons are extremely primitive in comparison with the complexity of biological neurons. So creating 100,000 million artificial neurons, for example, wouldn't mean that the resulting silicon brain would have the same capabilities as a human brain.

It took millions of years of evolution to make the brain of a biological cat; but with help of the CBM, De Garis thinks he should be able to produce something cat-like in a couple of years. The CBM does this through 'evolutionary engineering', a process of digital Darwinism by which only the fittest circuits for specific tasks survive.

Robokitty will need to see, for example, so its brain will require neural networks designed for pattern recognition like those described in the chapter on 'Sight'. Rather than laboriously programming these vision circuits by hand, De Garis has the CBM generate a random population of circuits and exposes them to identical inputs – in this case, things such as faces, household objects, straight lines and rounded surfaces. The CBM monitors the responses of all the circuits and measures how well each one performs. Most circuits don't respond at all, so the CBM discards these and uses the ones with the best pattern-recognition responses to breed the next generation of circuits. Hundreds of generations are processed in about a second. The CBM's computational power is critical to the evolutionary approach, since time is of the essence in both biological and artificial evolution.

Once an iteration of the cycle is complete, the CBM starts from the beginning again, exposing a new, mutated generation to the same inputs, monitoring the responses and weeding out the underachievers. This process is repeated hundreds of times – if the circuit is robust, it survives, mutates, reconfigures and reproduces; if not, it dies – until the computer generates a population of neural networks that performs pattern-recognition tasks adequately. From this population, which takes a few seconds to evolve, the best-performing circuit is extracted and inserted as a component into Robokitty's brain.

According to Michael Korkin, the Russian-born president of Genobyte, the Boulder, Colorado, firm that designed and built the CBM, the CBM itself has some of the same qualities as the biological brain. 'We routinely remove large portions of hardware from the CBM and often don't observe any significant degradation in its performance,' he says. 'Also, we don't instruct the neural network how to evolve a circuit for a specific task – and often we don't have the slightest idea how it's done. We only tell it the desired result, what we expect the behaviour to be like.'

In the ball–kicking scenario, for example, Korkin and his colleagues don't provide step-by-step instructions as Aibo's creators do. They simply tell the CBM, 'We want the robot to approach the ball and kick it.' The CBM's genetic algorithms, strings of computer code that use trial and error to combine and mutate to create new circuits in a way analogous to the process of natural selection through mutations in DNA, then attack the problem to come up with the right circuits. If Aibo and Robokitty were placed next to one another and they both kicked a ball in an identical fashion, it might appear as if there was no difference between the two; but only Robokitty could be considered intelligent – in a limited sense, at least – because it kicked the ball without a human programmer telling it how to do it. Robokitty is much closer to the way biological cats produce behaviour, though both robotic pets would pass the duck test.

'The CBM's ability to evolve these networks makes it possible to assemble tens of thousands of them into humanly architected artificial brains,' De Garis says. 'These brains will control thousands of different behaviours to generate a rich repertoire of biologically based functionality.'

Robokitty's brain needs tens of thousands of these circuits because individual networks are required for everything the robot does, from seeing, hearing, smelling, tasting and touching to walking, running, leaping, purring and sleeping. So once the CBM has evolved the pattern-recognition circuits, work must begin on the thousands of other sensory abilities and behaviours possessed by biological cats. At some point all these task-specific networks must be welded together. The robot's vision system, for example, must be able to communicate through its brain with other parts of its body so that the creature can successfully navigate its environment; and herein lies the flaw in De Garis's plan.

While the CBM can evolve and design the individual networks by itself, programming the different behaviour modules to work together has to be done by hand. In biological brains, neuron ensembles dedicated to specific tasks communicate with one another through an intricate mesh of links. These interconnections evolve spontaneously, but in Robokitty they have to be assembled manually; and that is a daunting task. 'You can't hand-code millions of modules with thousands of millions of neurons,' De Garis concedes. He has yet to come up with a way round the impasse. This binding problem throws De Garis's approach back into the realm of traditional manual programming and engineering, exactly what he had hoped to avoid with the CBM. Though there's no solution in sight, some researchers argue that evolution in the real world – rather than in a super-computer – is needed. 'It may take the same amount of time to evolve a complex artificial brain as it took for the biological one,' Korkin says. 'A population of robots might have to be put in the real world to live and go through millions of generations in real time, not computer time, interacting with the real physical world.'

In the meantime, De Garis is enlisting an army of programmers from across the globe to start linking up Robokitty's neural networks. Given enough people and perseverance, he may well succeed in connecting a few hundred modules together to demonstrate complex behaviours, such as getting Robokitty to play with a ball of string; but De Garis admits that his goal of building a lifelike autonomous kitten-like robot, much less a god, will remain unachieved until the manual linking problem is solved.

Despite the formidable technical obstacles, De Garis is convinced that the CBM is the beginning of a technology that will eventually allow people to create artificial intellects, or 'artilects', with intelligence 'trillions of trillions of trillions of times above the human level'. As these machines appear and grow ever more powerful, he believes humanity will split into two ideologically opposed camps: the 'cosmists', who favour building artilects, and the 'terrans', who don't. 'The biggest ethical problem to be faced this century will be the issue of species dominance – whether humanity should build massively intelligent machines,' De Garis says. He even predicts 'an artilect war' between the cosmists and the terrans sometime late this century that will determine the fate of the human race. De Garis, a cosmist, is in no doubt about who – or more accurately, what – will prevail.

De Garis's apocalyptic vision seems more in the realm of science fiction than scientific fact, especially given that Robokitty is still little more than a bunch of isolated pattern-recognition circuits without a body. Something more than evolutionary engineering alone will be needed before artificial creatures become even mildly intelligent. De Garis is candid about not yet knowing what that extra something is.

His approach is simply to connect enough of the right kind of brain circuits together in the hope that intelligence will emerge. 'If you get enough quantitative change, you get a qualitative change,' De Garis says. This is, as Steve Grand argues, the way that biological brains seem to do it. Why shouldn't it work in silicon too?

One reason why not is suggested by Jack E. Steele, the man who coined the term 'bionics' back in 1958. 'I get kind of irritated with people who make a neuron and say, "Well, if we hook enough together it will think,"' he once remarked. 'Like, "Birds fly with feathers. Let's glue a bunch of feathers on a board and it will fly."' John Searle's dictum – that artificial brains 'must be able to cause what brains cause' – must still be observed. Just because Robokitty might one day be able to kick a ball, it doesn't necessarily follow that it has the mind of a biological cat. There's clearly still a long way to go before Robokitty crosses the threshold of consciousness. When it comes to mind, appearances can be deceiving.

'The essence of consciousness is that it consists in internal qualitative, subjective mental processes,' Searle writes in *The Mystery of Consciousness*. 'You don't guarantee the duplication of those processes by duplicating the observable external behavioural effects of those processes ... To try to create consciousness by creating a machine which behaves as if it were conscious is irrelevant, because the behaviour *by itself* is irrelevant. Behaviour is important to the study of consciousness only to the extent that we take the behaviour as an expression of – as an effect of – the inner conscious process.'

So the duck test may not be an accurate gauge for evaluating conscious robots, after all. It could be a very long time indeed before we need fear the risk of being displaced or enslaved by a race of massively intelligent machines. For the time being at least, we seem safe from becoming what the English novelist Samuel Butler called a 'parasite upon the machines, an affectionate machine-tickling aphid'.

MIND INTO MATTER

Tama certainly does enjoy a good tickle, but it's difficult to imagine it or its descendants taking over the world. Playing with Tama, a robotic pet cat curled up on a table in Takanori Shibata's lab in the mechanical engineering department of the Japanese Ministry of Economy, Trade and Industry in Tsukuba, just north of Tokyo, the chances of an artilect war seem remote.

Tama is a cute salt-and-pepper tabby that turns its head and fixes me with its impassive gaze when I call its name, then slowly blinks its eyes in that infuriating way cats have of conveying utter indifference and disdain. When I pick it up and scratch its head, though, it purrs and scrunches up its eyes in an amazingly realistic rendition of sensual pleasure. At this level of interaction the only thing that betrays Tama as something other than a biological cat is the fact that it's not soft to the touch. There's a lot of hardware and circuitry under its fake fur.

Tama, the result of a collaboration between Shibata and the Japanese industrial corporation Omron, can execute such realistic feline behaviours because it has senses that enable it to see, hear and feel what goes on around it and an 'artificial mind' that enables it to respond accordingly. The robot has four auditory sensors that allow it to localise sounds, as well as speech-recognition software so that it can recognise its own name and distinguish its master's voice from those of other people. It has CCD cameras for eyes that can track movement and recognise faces plus eight separate tactile sensors – on its back, head, chin and cheeks – so it can distinguish between an affectionate caress and an angry slap.

This rudimentary sensorium is complemented by about a dozen actuators that coordinate the robot's movements. Motors in the eyelids, neck, front legs and tail power all Tama's physical responses. When I graze a whisker, for example, the robot shies away using its front legs. When I slap it hard on top of its head, it scurries backwards. 'Tama is a machine that understands human beings and expresses its own intentions,' says Shibata. 'It's important for robots to respond interactively and behave like living things so that humans will establish emotional bonds with them.'

Emotional bonds are also important because Omron wants to market Tama as the first in a new range of sociable robots that will serve as household pets and companions for the elderly. After the commercial success of Aibo, Omron hopes Tama will appeal to cat-lovers who live in cramped city dwellings in which biological pets are not allowed, as well as to older people looking for the affection and companionship of a pet without the mess. Shibata also envisions applications in what he calls 'robot-assisted therapy', in which Tama and its robotic progeny serve as entertaining and non-infectious playmates for children requiring extended hospital stays.

Tama is a reasonably convincing cat, albeit within a very limited domain of interactions, because Shibata and his collaborators have provided it with an 'artificial mind', a neural network programmed with a set of basic instincts that motivate its internal states and actions. Because only its instincts are programmed and not its behaviours, Tama can be considered an early version of the kind of robotic kitten De Garis would like to build.

Tama's internal states – what we might describe as moods or dispositions in a biological pet – are determined by what happens when its inner drives and the outside world collide. The robot's artificial mind has two sets of neural nets: the first contains instructions for its built-in preferences and the second, called an 'emotion generator', controls its changing internal states, which in turn direct its behaviours. For example, in Tama's mind gentle stroking is always positive, or 'good', and rough hitting is always negative, or 'bad'. Tama is programmed to seek the good and avoid the bad, so when its tactile sensors detect gentle stroking, the emotion generator registers this stimulus as good and instructs the actuators in its eyes to scrunch up in satisfaction. When the tactile sensors detect a rough slap, on the other hand, the emotion generator registers this as bad and instructs the actuators in its legs to pedal backwards. So the robot's internal states, which range from anger and surprise to anxiety and satisfaction, change according to how it is treated.

Another of Tama's internal imperatives is to play. If the robot is left unstimulated for too long, it becomes restless and begins to seek attention by moving and purring. Once the robot finds a potential playmate, whether it be a person or a plastic ball, it trains its visual and auditory sensors on the object and attempts to engage it in interaction. If Tama is successful, and this particular object turns out to provide the needed stimulus, the robot stores that information in its neural network and seeks out the same object the next time it is bored; but after a period of continuous stimulation, Tama gets tired, and the emotion generator registers fatigue. So the robot becomes disinterested, stops playing and eventually nods off.

The difference between Tama's programmed and unprogrammed

behaviours may be invisible to the user, and from the outside, it might well seem trivial anyway; but it is precisely this difference that distinguishes 'artificial' artificial intelligence that's preprogrammed from 'real' artificial intelligence that is based on biology. 'As a result, people think the robot has a mind of its own because it can express emotion and generate its own behaviour,' Shibata says. 'Many people believe Tama has its own mental life.' But as far as Tama and Robokitty are concerned, the mind is most definitely in the eye of the beholder.

While Shibata has given Tama the basic drives to play and seek positive feedback, Tetsuya Ogata of the Brain Science Institute of Japan's Institute of Physical and Chemical Research has given his robot even more primitive instincts: self-preservation and survival. 'If a system has the survival instinct and can communicate, it's possible that mind can emerge,' Ogata says. WAMOEBA (the Waseda Artificial Mind On an Emotion BAse) is the system Ogata designed to test this hypothesis.

As we've already seen, there must be a body before there is a mind; and one of the mind's most salient features is its ability to perceive the body's changing conditions, perceptions based on information from the senses regarding the outside world and from the central nervous system regarding the internal world. In humans, changes in the body's internal states are caused by the endocrine system, a network of glands that regulate physical processes by secreting hormones that are carried through the bloodstream. The endocrine system is also responsible for maintaining the body in a state of homeostasis, the process by which a biological system keeps itself fit while adjusting to variable external conditions. WAMOEBA has the full complement of silicon senses we've encountered in the other robots, but it has something they don't have: its own model of a biological endocrine system.

WAMOEBA is roughly humanoid in shape and about the size of an eight-year-old child. It looks a bit like the robotic maid in the 1970s cartoon series *The Jetsons* and, like that character, moves itself around on its own wheels. Humans can communicate with WAMOEBA by talking to it, touching it, or by waving or clapping the hands. The robot responds by approaching or retreating, stretching out its arms as if to embrace, making eye contact and speaking. Its tone of voice becomes excited when it perceives something interesting, and it wails in a plaintive tone when it is 'bored' from understimulation.

In human beings homeostasis is maintained by balancing things such as muscle tension and body temperature. WAMOEBA's version of the endocrine system manages the robot's electricity consumption and the temperatures of its motors and circuits. WAMOEBA strives to maintain its energy levels and temperature constant since its survival depends on keeping these internal conditions in a healthy state: too little electricity and the robot's battery runs down; too much heat in its circuitry and it blows a fuse.

Maintaining a constant temperature is relatively easy for WAMOEBA. When it detects any of its circuits or actuators overheating, it independently activates its cooling fans. If that's not enough, the robot will shut itself down until its internal temperature returns to an optimal level. It does this without any specific programming. Just as our endocrine system cools us down after a workout without us having to consciously think about it, WAMOEBA's endocrine system influences its neural networks to cool itself down. Similarly, when the robot detects that the voltage of its battery is low, it becomes 'hungry' and starts foraging for food. To restore the proper balance in its energy level, WAMOEBA heads for the nearest electrical outlet, plugs itself in and recharges its battery.

WAMOEBA has no face, but it expresses its internal states by means of a touch screen on its torso. When homeostasis has been achieved and WAMOEBA feels 'good', the colour of the screen is yellow. When WAMOEBA's internal state is unsatisfactory, the screen turns red. When the robot is calm, 'sad' or neutral, the colour is blue; but WAMOEBA displays these basic colours only rarely. Most of the time the robot's internal condition is so complex that it displays a mixture of these three colours, indicating that its inner state is a mixture of satisfaction and dissatisfaction, stability and instability.

In one study that was modelled on Pavlov's experiments with salivating dogs, Ogata conditioned WAMOEBA to display a yellow colour every time it heard a particular tone. First WAMOEBA was exposed to the tone each time it recharged its battery. While charging, WAMOEBA produced the yellow colour on its display. Eventually, the robot came to associate the tone alone with the 'pleasant' sensation of recharging, so the yellow colour appeared when WAMOEBA heard the tone even when it wasn't recharging. This experiment showed that even an artificial endocrine system can undergo real conditioning. 'The purpose of this research is to design a robot in which intelligence and emotion equal to that of humans can emerge,' Ogata says, 'to achieve a kind of human–robot symbiosis. To achieve that, the robot has to make the observer feel that it has a mind, since that's the basis for all emotional communication.'

One of the most successful robots at making the observer feel that it has a mind is Kismet, an expressive robotic head developed by Rodney Brooks and his colleagues at MIT's Artificial Intelligence Lab. Though Kismet is just a head that looks a lot like a mischievous elf, it has enough sensory and motor abilities to communicate with humans in what seems to be a genuinely emotional way.

Brooks is the key theorist behind the embodied artificial intelligence movement. Since mind and body go together, he reasons, what kind of mind a robot has is likely to depend on what kind of body it is given. If you build an insect robot, you're likely to get an insect-like mind; if you build a humanoid robot, you're likely to get a humanoid-like mind. Brooks is interested in the latter. 'If we are to build a robot with human-like intelligence, then it must have a human-like body in order to be able to develop similar sorts of representations,' he has written. 'Intelligence cannot be separated from the subjective experience of a body.' To replicate human-like intelligence, Brooks provided Kismet with a mechanical version of a human head and engaged it in one of the most primal and emotional of all human interactions: the relationship between an infant and a parent.

Like the other robots discussed in this chapter, Kismet is programmed only with a set of basic drives: a social drive, which impels the robot to interact with people; a stimulation drive, which motivates the robot to seek stimulation from its environment; and a fatigue drive, which directs the robot to rest after an intensive round of stimulation. The behaviours that result from these drives are not preprogrammed, but emerge as a result of the robot's interactions with the environment.

Like a human infant, Kismet started life in a rather helpless state and picked up its social and communication skills through interaction with its care-givers. This interaction is typical of a baby–parent relationship: lots of high-pitched babbling and exaggerated facial expressions. Building on these early learning experiences, Kismet is now capable of producing much more sophisticated and diverse behaviours; but Kismet is still very much a child in terms of intelligence: its range of competence is fairly limited and, like a human infant, when any of the robot's needs isn't met, it lets you know about it in no uncertain terms.

Kismet's social drive motivates it to seek out people, so when it detects a human in the immediate vicinity its eyes widen, it wiggles its ears and starts a playful conversation. Should the interlocutor move out of the robot's visual range, Kismet's eyelids droop and it displays a sad expression. Four actuators at the corners of its lips enable it to frown, or smile when the input from its sensors is more to its liking. Kismet can also prick up its ears to show interest or fold them back against its head like an angry dog. The robot can raise and lower each eyebrow independently, so it can furrow its brows in a show of frustration or cock one eyebrow in a display of Mr Spock-like surprise. It hears through microphones and speech-recognition software and has a voice synthesiser that makes it sound like a young child.

With such lifelike facial expressions and emotional displays, does Kismet have a mind? The answer is no – not yet, in any case. But it's not implausible to imagine the day when robots like Kismet will be recognised as having minds, when switching off such a machine might be regarded as an act of murder. Could a robot ever be given equal status to a person? Could it ever have a soul?

'Definitely!' affirms Anne Foerst, a Lutheran minister from Germany and director of MIT's God and Computers project, in which she worked with Kismet from its beginning in 1994 to explore the theological implications of artificial intelligence. 'Robots will have souls as soon as we confer personhood on them.'

Foerst prefers the term 'personhood' to mind, since personhood is ultimately what Kismet's creators are striving for. 'A newborn baby doesn't have a mind, consciousness or self-awareness,' Foerst argues. 'All these features develop only in social interactions between the infant and care-givers. If a being is not integrated into these processes, there will be no mind. Despite this, an infant is considered a person. So personhood does not depend on having a mind, but on the intrinsic value placed on the creature in question.'

In Hebrew theology, Foerst points out, the soul is considered an emergent phenomenon, something that comes into being only through the interaction of a person with God and with his or her community. Once robots are treated as valued beings and can physically participate in interactions among humans, Foerst suggests that there is no reason why they couldn't develop personhood, and therefore souls, as well. 'Humans are human because we are social,' Foerst says. 'The dignity of personhood is conferred inside a community. Our machines will only achieve this status when we can engage in social interactions with them and when we include them inside our communities. When we start treating our machines as if they have souls, they will start acquiring them.'

When Foerst began her work with Kismet, she admits she thought the tenets of artificial intelligence – one of which is that human beings are nothing more than very sophisticated machines – were just plain wrong. She found it impossible to reconcile that mechanistic view with her faith, which imputes to human beings a spiritual dimension that goes beyond the physical machinery of the body. Through her study of artificial intelligence, however, she realised that human beings are, in fact, just a collection of wonderful little mechanisms – genetic, chemical, neurobiological, physical and social. Foerst confesses,

> We are machines, but we don't treat each other like machines. We treat each other within a social framework with respect and dignity because we recognise that there's something special about us. We are machines, but we are also more than machines. This realisation was shocking at the beginning. I had always had faith in the one truth of theology. This truth was not shattered but it was certainly modified. My faith was enriched by science.

Ironically, the field of artificial intelligence itself is undergoing a belated realisation of the importance of the body, just as the Christian religion since Saint Augustine has struggled to overcome its rejection of the body as at best a distraction from, and at worst a hindrance to, the spiritual life. Artificial intelligence researchers went through a similar period of corporeal denial, focusing on simulating intelligence inside computers rather than inside bodies in the physical world. Through her work with Kismet, Foerst is exploring the importance of embodiment and social interaction for the eventual emergence of mind inside artificial life forms. If anything resembling a soul can emerge in a robot, then Foerst's relationship with Kismet is as good a place as any to start looking for it.

'We are creatures,' Foerst says. 'There's nothing special about us in that sense. Just because the spark of consciousness developed in us, there is no reason why it can't develop anywhere else. We can create robots as partners, just as God created us as His partners. The foundation of Christian faith is that God created us in His image. When we create conscious machines, we are participating in His creativity, celebrating it, glorifying it.'

In the 1950s Alan Turing, the English mathematician whose ideas were a major contribution to the development of the digital computer, addressed a similar concern in his essay 'Computing Machinery and Intelligence' when he wrote, 'In attempting to construct such [conscious] machines we should not be irreverently usurping His power of creating souls, any more than we are in the procreation of children: rather we are, in either case, instruments of His will providing mansions for the souls that He creates.' The house that Turing envisioned now has many mansions in which all kinds of beings have taken up residence, from robotic orang-utans to humanoids like Kismet. So far, none of these creatures has travelled far enough along the machine–life form continuum to be regarded as a person; but that is still the ultimate goal. 'If the spirit descends on our machines, who put it there?' asks Foerst. 'We don't know. All we can say is that it emerged.'

If the thought of a baby robot with a soul sounds disturbing, well, it is disturbing. The evolution of such creatures would be as drastic and culture-altering as the discovery that the earth revolves round the sun or that there was intelligent life on other planets. Norbert Wiener caught this sentiment when he wrote,

For the idea that God's supposed creation of man and the animals, the begetting of living beings according to their kind, and the possible reproduction of machines are all part of the same order of phenomena *is* emotionally disturbing, just as Darwin's speculations on evolution and the descent of man were disturbing. If it is an offense against our self-pride to be compared to an ape, we have now got pretty well over it; and it is an even greater offense to be compared to a machine.

But if robots like Lucy and Kismet one day do take on the mantle of personhood, we may just have to get over this offence, too.

THE ULTIMATE INTERFACE

'Brains and computers are different,' says Peter Fromherz, director of the Max Planck Institute for Biochemistry in Martinsried, near Munich. 'The problems to be solved are different, the processing of information is different and the materials have different physical and chemical natures.' Nevertheless, Fromherz has spent the past two decades trying to join these two very different things by linking neurons directly to computer chips. 'I try to unite what doesn't belong together,' he says.

Despite their dissimilarity, brains and chips are coming together in ever more unlikely unions. Researchers at Northwestern University Medical School in Chicago, led by Sandro Mussa-Ivaldi, have connected the brain of a sea lamprey, a kind of salt-water eel, to a two-wheeled robot. The lamprey brain, kept alive in a nutrient solution, responds to electrical signals from two optical sensors mounted on the robot. When the sensors detect light, signals are sent to the lamprey brain, which interprets them and instructs the robot's wheels to move either away from or toward the light. Like the robotic fish described in the chapter on 'Touch', this hybrid device is an important step in understanding how a biological nervous system can be integrated with a machine.

Such a mobile robot, minus the biology, was first built in the 1950s by British psychologist W. Grey Walter. The devices, described as 'electronic turtles', had very primitive sensors for eyes yet could navigate complex environments, find electrical outlets to recharge themselves unaided, and learn by association. In 1984, neuroscientist Valentino Braitenberg published *Vehicles*, a charming collection of thought experiments in which he suggested ways in which intelligent creatures could be constructed from simple robots with senses. These senses were very basic: a visual sensor caused the robot to approach or avoid a source of light; a tactile sensor caused it to avoid obstacles; an auditory sensor caused it to locate specific sounds. The vehicles' wiring – which included 'threshold devices', a kind of artificial synapse – became increasingly complex as the robots acquired more senses and the ability to perform more complex tasks. In the end, Braitenberg suggested that his vehicles would be able to perform acts indistinguishable from high-level mental functions such as learning, association, problem-solving, memory and thinking. Mussa-Ivaldi's hybrid robot is one of Braitenberg's vehicles come to life.

Fromherz is working to develop even more sophisticated neuron–silicon devices. He started linking neurons and chips back in 1985, working with giant Retzius cells from the ganglia of leeches. Though leeches are not normally known for their scintillating intellects, the Retzius cells were chosen because they are big – at 60 micrometres across, they are positively obese as far as neurons go – and therefore easier to manipulate under a microscope. Fromherz mounted the nerve cells on a specially designed chip and coaxed the Retzius cells to make connections with the chip's transistors. When the neurons were stimulated into firing, the chip registered the charge. Though it didn't look like much at the time, Fromherz succeeded in creating the first neuron-to-silicon junction.

In 1995 Fromherz developed a silicon-to-neuron junction so that the chip could talk back to the nerve cell: an electrical signal transmitted by the chip caused the leech neuron to fire, initiating the cascade of biochemical events that allows nerve cells to communicate with one another.

Since then Fromherz has moved on to bigger and better things, and slightly higher up the evolutionary ladder. Instead of leeches, he now works with the brain cells of snails and rats. In his most recent experiments he's managed to create a working synapse between two snail neurons mounted on a chip. The array consists of a silicon-to-neuron junction to stimulate the nerve cells, a working electrical synapse between the two neurons themselves, and a neuron-to-silicon junction so the chip can record the signals from the cells. When one cell is stimulated by the chip, it fires off a message to the other cell, which in turn sends an impulse back to the chip. The chip and the neuron talk to one another by batting electrical signals back and forth like ping-pong balls.

One of the main technical obstacles to integrating electronic circuitry with biological neurons is getting the nerve cells to make the proper connections, and devising a way to hold them in place so that they don't slip off the electrode array. To get around this problem, Fromherz is culturing the neurons right on the chips rather than trying to attach them later.

Again working with the humble snail, this time *Lymnaea stagnalis*, Fromherz has managed to control the formation of synapses on a chip. To do this Fromherz lays down a row of signal molecules along the narrow lanes of circuitry etched into the chip. Then, when the neuron is placed on the chip, its dendrites stretch along these lanes like a vine following the course of a flower bed. Working synapses form when the dendrites of two neurons collide in one of these narrow lanes. 'The dream system is an array of neurons with dendrites and axons growing in defined directions and with synapses formed at defined positions,' says Fromherz.

Such a system is important because it would permit the study of neural signalling in controlled environments rather than in the messy and hard-to-reach areas of living brains or random cell cultures; and in invertebrates such as snails, networks of just a few neurons can give rise to biological functions such as rhythmic synaptic firing. The main goal of such experiments, at this stage, at least, is to learn more about how neurons work rather than to devise some kind of silicon brain implant.

For now Fromherz's work is very basic, amounting to simply trying to understand the physics of the interface. How do you get a neuron to stick to a computer chip? How do you keep the cell moist and nourished? How do you get primitive signals to bounce back and forth between slime and silicon? So far he's managed to get two neurons conversing with each other and a computer chip. In a couple of years he hopes to have maybe a dozen neurons chattering away on chips; but any serious computation requires millions of neurons working in concert. No one yet knows how, or if, this many cells can be connected to a chip; and even if they can, the language of the human brain is still so poorly understood that if this many cells were talking with chips, no one would know what they were saying. Fromherz says:

> These experiments show how large a gap really exists between test-tube experiments involving a neuron–silicon junction and the inter-facing of the brain and the computer. As long as we do not know how

the brain functions, an informational interface remains impossible – even if the physical junction were feasible. To speculate now about things like brain prosthetics and mind–machine melding is like the ancient Greeks speculating about atomic bombs: it's possible, but it's a long way off.

Keiichi Torimitsu, leader of the molecular and bioscience research group at Japanese telecommunications firm NTT in Atsugi, Japan, thinks brain prosthetics are a long way off too, but he's already demonstrated simple information processing by neurons cultivated on a computer chip. Torimitsu took a group of neurons from the cortex of rat embryos and grew them on a sixty-four-channel electrode array bathed in nutrient solution. As the neurons crept across the array like moss, junctions formed between the biological and silicon synapses. After about four days the neuron–silicon synapses began to show some firing activity. After about two weeks, many synapses displayed the synchronised firing activity common in the rat cortex itself. Like Fromherz, Torimitsu is looking for clues about how the human brain works. 'By watching the neuron–electrode interface at work, we hope to understand how the brain processes information and hopefully learn how to control those processes,' he says.

In one experiment, Torimitsu and a colleague took neurons from a rat retina, part of the rat's brain called the thalamus that processes visual signals, and part of the rat's visual cortex that interprets what it sees. He cultured these neurons on a single chip and watched as the neurons grew and formed connections with each other and with the electrode array. In the rat brain, visual signals proceed from the retina to the thalamus to the visual cortex, and Torimitsu wanted to see if the same flow of information would take place on the chip. When the neurons were stimulated with electrical signals from the array, the electrodes recorded signals that proceeded in the same way as they would have in the brain: first the retinal neurons responded, then those from the thalamus and finally those from the visual cortex. This suggests that a hybrid brain–computer interface could process information in much the same way as neurons alone, opening up the possibility of neural prosthetics such as the 'brain phone', a mobile communications device that Torimitsu suggests would be all in your head. 'Once we understand in more detail how information is pro-

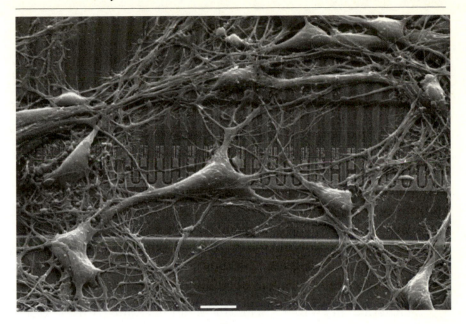

Figure 13. Rat neurons growing on a silicon chip.

cessed in the brain, we will be able to build an effective interface between chips and neurons,' he says. One effective interface between chips and neurons is already in use as a treatment for Parkinson's disease.

In neurodegenerative disorders like Parkinson's and Huntington's diseases, the release of neurotransmitters is disrupted in the brain, causing uncontrollable physical movements. In Parkinson's this results in the patient experiencing tremors, rigidity and a general slowness of movement. Chip-assisted neurons like those being created by Fromherz and Torimitsu might one day provide a treatment for this condition. Implanted electrode–neuron arrays could take over functions in damaged areas of the brain or be used to monitor neurotransmitter release, alerting physicians when intervention might be necessary.

In the 1940s and 1950s researchers discovered that when a part of the deep brain called the thalamus was stimulated with electrical impulses, the tremors so characteristic of Parkinson's were temporarily reduced or eliminated. In 1995, surgeons began implanting electrodes in the thalamus of Parkinson's patients as a way for them to control their symptoms. The electrodes are connected to a stimulator unit placed in the patient's chest below the collarbone by an insulated wire that passes under the skin of the head, neck and shoulder. (The arrangement is similar to the one described in the chapter on 'Touch' that allows Brian Holgersen to move his hand.) When the patient experiences tremors or shaking, he holds a magnet over the implanted stimulator. The magnet causes the stimulator to transmit signals to the electrodes in the thalamus, which blocks the brain signals causing the tremor. Thanks to this implant many patients experience enormous improvements in their ability to do things such as write, eat, tie their shoes and brush their teeth.

Michael Holman, a fifty-six-year-old British journalist, is one of them. 'I feel liberated from a ghastly prison in which I have lived for sixteen years,' he says of the eleven-hour surgical procedure that left him with an electrode array in his thalamus and renewed hope for a more normal life.

Holman's surgery was supervised by professors Alim-Louis Benabid and Pierre Pollak of the Joseph Fourier University Hospital in Grenoble, France, the two men who pioneered the implant technique. Holman had four holes drilled into his head, into which titanium bolts were inserted to hold his skull immobile during the delicate operation. The surgeons used brain scans to control a robot probe that guided the electrodes into place in the thalamus. Holman was conscious during the entire procedure – fortunately, there are no pain receptors in the brain so this part of the operation did not hurt – so he could tell the surgeons when he felt his symptoms subsiding. This is how the physicians knew they had reached the right spot in the brain.

For Holman, the highlight of the operation – if there can be a highlight to such a gruelling procedure – was listening to the sound of the neural traffic that the electrodes picked up in his brain. 'It was the most extraordinary experience I've ever had,' he says. 'It was an ocean of sound, a huge river in full flood, awesome and inspiring. It felt like the essence of me but also the sound of mankind, as close to a religious experience as I have ever had.'

The benefits of the surgery were not all spiritual, however. 'The shakes and tremors haven't entirely disappeared, but they are reduced and for the first time in ten years I can sleep at night,' he says. 'My dyskinesia [the involuntary twisting and squirming movements that are a side effect of some Parkinson's drugs] has stopped completely. My joints are not stiff, my muscles are not cramped, and I am no longer taking as many drugs.' Holman is not cured, though. He still has Parkinson's. But he does have his essence back, something that the disease had robbed him of.

A TRIP DOWN MEMORY'S LANES

In his book *Kinds of Minds*, philosopher Daniel Dennett suggests that one explanation for the greater intelligence of human beings is not our larger brains but 'our habit of *off-loading* as much as possible of our cognitive tasks into the environment itself – extruding our minds ... into the surrounding world, where a host of peripheral devices we construct can store, process, and re-represent our meanings, streamlining, enhancing, and protecting the processes of trans-formation that *are* our thinking. This widespread practice of off-loading releases us from the limitations of our animal brains.' It also releases us from the burden of having to remember too much, and memory is a key area in which the brain and computer chips intersect.

In his *Confessions*, Saint Augustine marvelled that the

faculty of memory is a great one, O my God, exceedingly great, a vast, infinite recess. Who can plumb its depth? This is a faculty of my mind, belonging to my nature, yet I cannot myself comprehend all that I am. Is the mind, then, too narrow to grasp itself, forcing us to ask where that part of it is which it is incapable of grasping? Is it outside the mind, not inside? How can the mind not compass it? Enormous wonder wells up within me when I think of this, and I am dumbfounded.

In many ways, neurobiologists (not to mention theologians) are still dumbfounded by memory's vast, infinite recesses. Recollections of the past were once believed to be enclosed in compact bands of neurons called 'engrams'. These putative biochemical time capsules were thought to lie dormant in the brain until retrieved from storage by some present association. One engram, for example, might contain all the sensory and emotional data associated with your first kiss;

another all the information about what exactly you were doing when the Berlin Wall came down on 10 November 1989. In the early 1980s, some researchers even claimed to have identified individual proteins in worms that encoded memories for specific learned behaviours. Transfer this memory molecule from one worm to another, they proposed, and you transfer the memory too.

Neither the engram nor the memory molecule theory held up under closer scientific scrutiny. The engram concept proved itself too static a model for how the brain in general – and memory, in particular – works. Research with positron emission tomography (PET) imaging, for example, has shown that many different brain regions, rather than specific clusters of cells, are at work in even the simplest acts of recollection. The existence of memory molecules, on the other hand, is unlikely for purely practical reasons. If each memory had its own protein, an individual's brain (or even a worm's, for that matter) would accumulate some 100 kilos of memory proteins – more than the weight of an average person – during the course of a normal lifetime.

Eric Kandel of the Howard Hughes Medical Institute at Columbia University's Center for Neurobiology and Behaviour has taken some of the mystery, though none of the wonder, out of memory through his work on long-term potentiation, one of the mechanisms by which recollections are believed to be stored in the brain. Working with mice and the snail *Aplysia*, Kandel has shown that memory formation appears to take place in two phases: short–term memories that last a few minutes followed by long-term memories that last a few days, a few years or an entire lifetime. Long-term potentiation is the increase in synaptic strength that occurs among a specific cluster of neurons when new information is learned, remembered and stored in the brain.

Kandel's research suggests that a region of the brain called the hippocampus is crucial for the formation of long-term memories. People with damage to this part of their brain have no lasting powers of recall; they can be introduced to someone at a party, for example, and a few minutes later, when presented with the same person, claim they have never seen them before. Kandel discovered that in order for a memory to become firmly lodged in the brain the connections among the synapses that fire in response to a given stimulus – in this case, say, a particular face – must become mutually reinforced.

If you meet someone only once at a party, the synaptic pattern that encodes this encounter will have only been weakly stimulated. If you see them again a few minutes later, you'll surely recognise them; if you see them again several years later, though, you may not. If, however, you meet the same person day after day, your synapses will have ample opportunity for this particular pattern of stimulation to become strongly reinforced; and if you see this same person again after an absence of several years, you'll still remember who they are. On the neuronal level, this process is called long-term potentiation and is believed to be responsible for all our most durable recollections, from recognising faces to remembering how to ride a bicycle or play the violin.

Kandel's work and that of others has shown that memories are not to be found in any single molecule or neural group, but in an intricate and ever-shifting net of firing neurons and crackling synapses distributed throughout the brain. Memory is not, as was previously thought, some vast cerebral warehouse filled with rows and rows of neatly ordered filing cabinets. It is rather more like a labyrinth, the twistings and turnings of which rearrange themselves completely each time something new is experienced and later recalled.

When we remember, two complementary processes take place: we learn something new and later recollect the experience. The actual learning initiates a cascade of protein synthesis and neural activity in the brain, which eventually leads to the formation of a short-term memory. These memories, which are unstable and easily disrupted, take shape when the cerebral cortex is flooded with the neuro-transmitter glutamate, the primary information-carrier in the brain. Glutamate opens the communication channels between neurons so that information about the experience, encoded in electrical and biochemical signals, can be transmitted through synapses, the crucial first step in making memories.

While weak and rapidly decaying memories evoke only a single wave of glutamate synthesis, the formation of long-term memories requires a second wave. Crucial to this phase is the production of a class of proteins known as cell-adhesion molecules. As the name implies, cell-adhesion molecules have sticky ends, arrays of Velcro-like protrusions that enable them to cling to the sticky ends of their partner molecules. As the newly minted cell-adhesion molecules latch on to one another, the arrangement of synapses in the brain changes. Like a river whose tributaries are constantly flowing into new configurations, the skein of synaptic connections shifts as the memory becomes embedded in the brain. The memory of the new experience is caught in this synaptic lattice like a fly in a spider's web. In neurobiological terms, this pattern of synaptic change is the memory itself.

If memories are indeed held in the brain as fluctuating patterns of synaptic connections, then how are they recalled? How does a childhood memory − of reading your first word, f-o-x, when you were six − persist for a lifetime, while the name of someone you just met vanishes within minutes? Why did the taste of a morsel of cake dipped in tea remind Marcel Proust of his boyhood Sunday mornings with Aunt Leonie? And how does memory endow us with a sense of self and personal identity?

According to Portuguese-born neurologist Antonio Damasio, head of the department of neurology at the University of Iowa College of Medicine, remembering is neither a tranquil nor a passive process: assembly is required. The brain is not a computer, Damasio argues, so recall is not simply a matter of clicking the proper icon to call up the desired document from the brain's hard disk. Memories must literally be re-membered, put together again from pieces found in various parts of the brain. 'The brain does not file Polaroid pictures,' Damasio says. 'Memory depends on several brain systems working in concert across many levels of neural organisation.'

In the short story 'Funes the Memorious', the Argentine writer Jorge Luis Borges relates the fate of a young man who, after a fall from a horse, is unable to forget any experience, whether real or imagined, from the present or the past. 'His perception and his memory were infallible,' Borges writes. 'He knew by heart the forms of the southern clouds at dawn on the 30th of April, 1882, and could compare them in his memory with the mottled streaks on a book in Spanish binding he had only seen once.' For poor Funes, the present was almost intolerable in its richness and sharpness, as were his most distant and trivial memories.

Fortunately for most of us, the neural organisation for memory in our brains is not nearly as implacable as that of Funes. Memory is, in fact, a notoriously faulty and unreliable faculty. We tend to perceive and remember only what we consider novel or important. The rest is indiscriminately dumped and forgotten. We have such selective recollection, Damasio suggests, because memories were developed more to predict the future than to retrieve the past. 'Our brains evolved to help us navigate the world safely,' he says, 'recognising and remembering danger so that we could avoid it in the future. For that purpose, a few key perceptions are enough.' Spot a twisted shape in the water and you jump, even though closer examination reveals it to be only a submerged stick and not a poisonous snake. 'If the brain stored every perception in complete detail,' Damasio says, 'there would be an explosive circuit overload.'

What we do remember, Damasio believes, is recreated in the brain through convergence zones, neurobiological crossroads where some of memory's many strands briefly meet to conjure up images of the past. But rather than storing a complete record of all the physical and psychological data associated with a particular memory – Funes' recollection of the southern clouds at dawn, for example – convergence zones store only that information necessary to reassemble the approximate record from neural firing patterns held in reserve at other sites in the brain.

To call even the most mundane memory to mind involves compiling and collating enormous amounts of sensory and psychological information. When Funes watched the sun rise on that spring day in 1882, different regions of his brain were busy processing data regarding the colour of the clouds, the sound of any birds that may have been singing, the smells in the air, the emotions he was feeling and thoughts he was thinking, as well as thousands of other impressions and sensations. All of this was held for an instant as a pattern of ignited neurons and chattering synapses. 'But this massive load of data is never collected into one place for processing and interpretation,' Damasio explains. 'Instead, it remains distributed over a wide area of the brain.'

This is where the convergence zone comes in. When we recall a person or an event, the convergence zone serves as a kind of neurobiological instruction manual, directing the appropriate neurons in the appropriate brain regions to reassemble themselves in a firing pattern that approximates to that of the original experience. Funes' view of the southern clouds at dawn set off a chain reaction of neural firing in his brain. The shape of the clouds reminded him of a particular book, and the convergence zone dutifully triggered the recreation of a synaptic pattern that exactly corresponded to Funes' glimpse of a volume in mottled Spanish binding he saw only once at a completely different time and in a completely different place. The memories thus recreated are not accurate in every detail – except in Funes' case, of course. They are replications, not duplications, of the original event.

Convergence zones do not therefore orchestrate the full symphony of past experience. They provide the score, but it's up to the synaptic ensembles involved in each individual memory to make the music; and just as with any musical performance, it's impossible to play the same note exactly the same way twice. 'Whenever we recall a given object or experience,' Damasio says, 'we do not get an exact reproduction but an interpretation, a newly reconstructed version of the original.'

Ironically, the volatile nature of memory is paralleled by the body's own internal dynamics. The cells in the body are continually dying as new ones are being born. Blood cells, for example, have a lifespan of a mere 120 days; taste receptors live only about ten days. Though most of the cells in our bodies will be replaced many times over during the course of a lifetime, we still feel that we haven't changed. In this respect, our bodies are definitely not our selves.

However, what if all your memories were suddenly replaced with new and different ones? Would you still be yourself? Strange as it may seem, the work on memory shows that – on a neurobiological level, at least – this is exactly what happens in the brain.

Memory is a constant work-in-progress. When an object or experience is recalled, the neural pattern corresponding to that memory flashes through the brain as clearly and as quickly as a lightning bolt; but like lightning, it is as swiftly gone, and the next time that same event is remembered the pattern will be different, changed by a complex network of new associations and experiences. Despite these unstable foundations, we somehow still manage to construct a stable idea of personal identity from this welter of mercurial memories. 'The self is not a little person inside the brain,' Damasio says. 'It is a perpetually recreated neurobiological state, so continuously and consistently reconstructed that the owner never knows it is being remade.'

There is a parable from Buddhist lore that mirrors on a psychological level what scientists have discovered about memory on the neurobiological level. The Buddha once compared the self to the flame of a torch whirled so swiftly in the dark that it seemed to form one unbroken hoop of light. As anyone who has done the same with a flashlight can attest, this continuity is an optical illusion created in the same way as we perceive the still frames of a film to be one seamless sweep of motion.

Memories are very much like the flame of that torch: fickle, inconstant, flickering. To our imperfect senses, though, they seem to form a coherent whole. Memories, however, are not fixed and immovable facts; they emerge from an ever-changing maze of neural firing formations and synaptic connections. They are in many ways fabrications, perpetually remade and replaced from raw materials that are in a constant state of flux.

TOTAL RECALL

Computer scientists are now bringing infallible computer memory to the aid of fallible human memory. In today's increasingly competitive marketplace, what struggling medical student or ambitious young lawyer wouldn't welcome the edge a memory aid could offer? For that matter, what about overworked air-traffic controllers, harried taxi drivers and aspiring actors, whose livelihoods depend on being able to flawlessly recall large quantities of information? The memory prosthesis is designed to extend the limits of the human brain by taking large chunks of information and recollecting them in the environment – in this case, a wearable computer.

These memory prosthetics incorporate many of the sensory technologies we've encountered in this book: eyes the better to see with and ears the better to hear with. Imagine coming home from work exhausted and harried and inadvertently dropping your car keys in the kitchen sink. A video camera in your lapel like the one described in the chapter on 'Touch' takes note and remembers. Twenty minutes later, as you're cursing under your breath because you can't find the keys and you're going to be late for your son's violin lesson, the computer gives a gentle aural or visual prompt: 'Try the kitchen sink.' In the age of information overload and attention deficit syndrome, the best place for many memories may be, as Dennett suggests, outside the mind.

A better memory, like thinner thighs or higher cheekbones, is one of those coveted personal attributes that is in constant demand. Newspapers and magazines regularly feature advertisements promising to reveal how you can improve your memory in a single evening. One such ad relates the 'amazing experience' of a satisfied customer who, thanks to a new mnemonic technique, was able to 'memorise the 2385 paragraphs of the German civil code!' This feat is presented as an enviable achievement, but the German civil code seems to me an awful thing to have cluttering up your cranium; and what the ad fails to mention is that this 'new' technique is actually a watered-down version of the ancient art of memory, a method that helped medieval orators and mystics remember speeches and spiritual arcana by associating texts with mental images arranged inside a 'memory theatre'.

For those not interested in taking the trouble to visualise medieval architecture, there's always the technological shortcut. Like so much 'new' technology, the memory prosthesis has a long and distinguished history. The portable devices appearing today trace their roots back to an article by the American inventor and engineer Vannevar Bush in the American magazine *Atlantic Monthly* in 1945.

In his piece 'As We May Think', written just after the end of World War II, Bush praised the technological accomplishments of science but noted a serious lacuna. 'For years inventions have extended man's physical power rather than the powers of his mind,' he wrote. 'Trip hammers that multiply the fists, microscopes that sharpen the eye, and engines of destruction and detection are new results, but not the end results, of modern science ... The summation of human experience is being expanded at a prodigious rate, and the means we use for threading through the consequent maze to the momentarily important item is the same as was used in the days of square-rigged ships.' Bush's solution: the memex, a mechanised way to navigate the information maze and augment the human memory.

Bush recognised that the human brain is not a computer. It does not proceed in relentlessly linear fashion but in fits and starts, making swift and nimble connections between seemingly unrelated things. The brain jumps back and forth in time, makes leaps of faith, inspired guesses, serendipitous digressions and brilliant mistakes. 'The human mind ... operates by association,' he wrote. 'With one item in its grasp, it snaps instantly to the next that is suggested by the association of thoughts, in accordance with some intricate web of trails carried by the cells of the brain ... Trails that are not frequently followed are prone to fade, items are not fully permanent, memory is transitory. Yet the speed of action, the intricacy of trails, the details of mental pictures, is awe-inspiring beyond all else in nature.' Like Saint Augustine, Bush was dumbfounded by memory's majesty.

Writing some forty years before the advent of the World Wide Web, Bush's musings on the mind parallel those of Tim Berners-Lee, the man who has done most to make the Web a reality. In his memoir *Weaving the Web*, Berners-Lee recalls an incident from his childhood that set him thinking about how computers could be made to forge spontaneous links like the brain. Berners-Lee writes:

> One day when I came home from high school, I found my father working on a speech for Basil de Ferranti. He was reading books on the brain, looking for clues about how to make a computer intuitive, able to complete connections as the brain did. We discussed the point; then my father went on to his speech and I went on to my homework. But the idea stayed with me that computers could become much more powerful if they could be programmed to link otherwise unconnected information.

Inspired by this desire to make computers more like human brains, Berners-Lee invented the Web; Vannevar Bush invented the memex.

The memex consisted of an ordinary desk equipped with two slanting, translucent screens on which images and texts could be projected, a keyboard for entering data and − this being 1945 − a complex array of buttons and levers for operating the device. The memex was intended to be 'a sort of mechanised private file and library' in which books, records, newspapers and periodicals as well as pictures and correspondence were stored on microfilm. Embedded in the desktop itself was a glass plate through which handwritten notes, memoranda and images could be photographed and entered into the memex. This complicated and bulky machine was meant to be the first step toward 'associative indexing, the basic idea of which is a provision whereby any item may be caused at will to select immediately and automatically another'.

Items are tied together in the memex through a manual and somewhat tedious process. When a user wants to build a new trail, he takes the two items to be joined – which are projected on the adjacent viewing screens – and assigns each one a code that corresponds to the other. Then, with the tap of a single key, the items become inextricably linked. In future when one of these items is recalled from the memex, the other can be summoned up by clicking a button below the corresponding code space. Bush wrote,

> When numerous items have been thus joined together to form a trail, they can be reviewed in turn, rapidly or slowly, by deflecting a lever . . . It is exactly as though the physical items had been gathered together from widely separated sources and bound together to form a new book . . . One cannot hope thus to equal the speed and flexibility with which the mind follows an associative trail, but it should be possible to beat the mind decisively in regard to the permanence and clarity of the items resurrected from storage.

A memex never forgets.

Bush envisioned the memex as a way for individuals to get quick and efficient access to the rapidly proliferating sum of human knowledge. In so doing, he anticipated such developments as hypertext, online databases and search engines.

> Wholly new forms of encyclopedias will appear, ready made with a mesh of associative trails running through them . . . The physician, puzzled by a patient's reactions, strikes the trail established in studying an earlier similar case, and runs rapidly through analogous case histories, with side references to the classics for the pertinent anatomy and histology . . . There is a new profession of trail blazers, those who find delight in the task of establishing useful trails through the enormous mass of the common record.

Bush even speculated about the development of portable devices that would allow people to add their personal experiences to the memex's mass of published material. 'The camera hound of the future wears on his forehead a lump a little larger than a walnut,' Bush imagined. 'It takes pictures three millimetres square, later to be projected or enlarged ... The cord which trips its shutter may reach down a man's sleeve within easy reach of his fingers.'

That 'camera hound of the future' sits on a leather couch on the third floor of the MIT Media Lab building in Cambridge, Massachusetts. Bradley Rhodes, a recent graduate, isn't wearing a walnut on his forehead but he does, as he himself describes it, have 'a big, bulky harmonica case stuck to my hat'.

The harmonica case, a Private Eye made by Reflection Technology, is positioned just above Rhodes' right eyebrow and holds a computer display, which is connected by means of a cable that snakes through Rhodes' beret and down his back to a small computer he carries in a satchel slung over his shoulder. The fingers of Rhodes' left hand tap furiously across a small hand-held keyboard called a Twiddler. As we talk, he keys in my name, the title of this book and other notes. Rhodes calls this contraption, which weighs about the same as a couple of good-sized textbooks, a 'remembrance agent'. It is, in effect, a portable memex.

During our conversation Rhodes' eyes frequently dart up for a quick glance into his harmonica case. As he enters keywords from our chat into the computer, the remembrance agent calls up files relevant to what we're talking about and projects them onto the display.

Wearable computers? Up pops a slew of e-mails between Rhodes and former Media Lab classmate Thad Starner. The relationship between computer memory and human memory? Up pops a link to Rhodes' doctoral thesis, which he later prints out for me. Favourite movies? Up pop reviews of *Inspector Gadget*.

'A personal digital assistant is blind, deaf and dumb,' says Rhodes. 'It can't see the environment, and it can't do anything but what I tell it to do.' The remembrance agent, though, is a search engine with senses, a suite of software applications and sensors (a video camera to look over your shoulder, microphones to catch conversations) that continually monitors the wearer's environment. By tapping into its own database as well as the Web, the remembrance agent is able to present information based on the wearer's immediate location, context and needs. The device is designed to be unobtrusive; it runs in the background and only alerts you when it's found something of interest. Apart from the harmonica case, no one would ever know you're using it.

The system Rhodes uses has just a few sensors, the most sophisticated being a radio-frequency receiver that can tell what room he's in by picking up beacon signals placed around the lab. Others, however, are integrating additional sensors into the device, including microphones equipped with speech-recognition software to retrieve documents based on conversation and cameras to determine what kind of environment the user is in: at work, for example, or at the shop.

'Human memory and attention are expensive,' Rhodes says. 'With the glut of information we receive every day, the remembrance agent is an aid to associative memory.' Rhodes uses his remembrance agent to record snippets from important meetings, archive e-mails (he hasn't deleted an e-mail in five years, he says) and as a kind of interactive diary. All his data, whether trivial or top secret, are poured into it. Without his remembrance agent, for example, Rhodes might not have remembered that his e-mail exchange with Thad Starner contained information relevant to this book.

The remembrance agent has a range of other potential applications. For simple reminders of appointments or things-to-do, it could whisper, 'It's time to go to the dentist,' or, 'Don't forget to call your mom.' For people who have trouble remembering names, an agent equipped with pattern-recognition software could help avoid embarrassing encounters. The next time you meet an acquaintance at a conference and go completely blank on his or her name, the device could scan the face, identify the person and call up vital stats such as occupation, interests of spouse and children as well as notes from the last meeting you had with this person. Remembrance-agent-like devices for people with neurodegenerative disorders like Alzheimer's that affect short- and long-term recall could wear 'memory specs' that sense patterns in the environment, such as faces, and provide helpful prompts – like 'This is your son-in-law, John' – to patients suffering from memory loss.

As the amount of information in the world increases, human attention will become an increasingly scarce – and therefore an increasingly valuable – commodity. A remembrance agent could help connect the dots between the constantly proliferating bits of isolated information; and it has one big advantage over the memex: the links are made by the computer automatically, without the need for manual coding. 'Intelligence doesn't stop at the skin,' Rhodes says. 'The outside world is a major part of our intelligence.' And the remembrance agent is another way to off-load our cognitive tasks onto that environment.

THE UNDISCOVERED COUNTRY

My father died twice. The first time was in an ambulance as he was rushed to the hospital after suffering a heart attack. His heart had stopped beating for several minutes before the paramedics were able to revive him. Afterwards, he described a vision he had had as he lay without a pulse in the speeding ambulance. He saw his brother Joe, who had passed away some years earlier, dressed in flowing white robes, sitting on a kind of throne at the top of a hill. As my father approached, Joe was laughing uproariously and waving his hand, urging him to turn around and go back.

The second time my father died was about three years later. Again, it was his heart that failed him. This time, though, his brother Joe wasn't there to wave him back to this world.

What my father encountered in the back of that ambulance was a classic example of a near-death experience (NDE). According to statistics compiled by the International Association for Near-Death Studies, some 35–40 per cent of people who have had a close brush with death later report an NDE. They commonly tell of a feeling that the 'self' has left the body, a sensation of moving through a dark space or tunnel, a vision of golden or white light, and the receipt in some form of the message 'your time has not yet come'.

Are NDEs evidence for life after death, or are they just the last, desperate projections of the dying brain? While science has offered explanations for why we age, it tells us very little about why we die, much less about what happens, if anything, after death.

Many people don't require hard evidence about what's on the other side: they are satisfied with the explanations provided for millennia by the world's religions; but for those not blessed with religious faith, science may never be able to prove whether the NDE is a newsflash from the hereafter or just one of the mind's grandest illusions.

David Darling, an astronomer and physicist who has written several books on death and dying, suggests there are two events that approximate to death: the near-death experience itself and the memory loss suffered by victims of Alzheimer's disease and traumatic brain injury. 'Losing your memory is perhaps as close as we can come to death without actually dying,' he says. 'The person is dead because the memories are gone.' Memories are the only things that give meaning and identity to our lives.

In his book *Design for Dying*, completed just before he died from prostate cancer, psychologist and LSD guru Timothy Leary explored various strategies for achieving immortality. One of the most promising tactics, he suggested, is to transfer an individual's consciousness into a supercomputer before death. 'If you want to immortalize your consciousness,' he wrote, 'record and digitize.'

What is a life, after all, but a lifetime's worth of memories and sensory impressions? Using the technologies described in this book, it would not be very difficult to record every moment of your life. A startlecam could capture your most precious and most tedious moments; a modified cochlear implant could record crucial conversations, ambient sounds and other auditory input; smell and taste sensors could store the recipes for all your most memorable meals and aromas; a GPS embedded in your clothing could track your every move while bio-sensors record every heartbeat, every breath; video tapes from conventional security systems could be compiled to depict you buying bread, leaving a bank, walking down the high street; even the brain signals that control your body movements could be recorded by a neural implant and played back later to control a robot that mimics the way you walk and talk. Download it all onto a computer and the result is a real-life ghost in the machine. Endowed with all your memories, this new computerised 'you' could be called up over the Internet to chat and interact with people just as you would – except that it would be doing so long after you had died. As Norbert

Wiener speculated back in 1964, it is conceptually possible for a human being to be sent over a telephone line.

If the difference between life and death is a question of memory, could death's sting be parried by replacing the fragile human memory with a more robust computerised one? Max More, president of Extropy Institute, which explores immortality-through-technology strategies, thinks so. Since memory is encoded in the fluctuating synaptic connections in the brain, the most promising way to achieve, if not immortality, at least 'superlongevity' is to transfer those evanescent patterns into a computer that can reproduce them at will.

Nanotechnology, the technique by which cell-sized, self-replicating robots construct atom-sized machines inside the human body, is the crucial missing link in all of the Extropian superlongevity schemes. The Extropians suggest that 'nanochips', artificial nerve cells the same size as biological nerve cells, might gradually replace the brain's own neurons. These nanochips, perhaps highly refined and miniaturised versions of the chips Fromherz and Torimitsu are working with, would hook up with a neuron, monitor its activity and then eventually assume its functions. In this way the chip would retain all the connections and information-processing capabilities of the biological neuron that it replaced, but with none of its vulnerability or mortality.

Once a sufficient number of chips was in place, the brain thus constructed would no longer be dependent on biology. It could become a body-hopper, taking up residence in other biological bodies, in non-biological organisms, or even on the Internet itself. 'We don't yet understand the brain, so we couldn't possibly upload it whole onto the Internet,' More explains. 'But we could gradually replace parts of the brain with synthetic components so that eventually the biological brain would become a vestigial organ. We wouldn't have to rely on biological processes or a single body. We would still need a body, but not necessarily one body.'

Roboticist Hans Moravec put the same proposition like this: 'Body-identity assumes that a person is defined by the stuff of which a human body is made. Only by maintaining continuity of body stuff can we preserve an individual person. Pattern-identity, conversely, defines the essence of a person ... as the *pattern* and *process* going on in my head and body, not the machinery supporting that process. If the process is preserved, I am preserved. The rest is mere jelly.'

Another Extropian plan for superlongevity involves cutting the brain into slices and preserving them in liquid nitrogen. Each of these slices would be scanned by a computer, which would reconstruct that person's synaptic connections so that its patterns are replicated inside an artificial brain. When the brain scan copy is activated that person, hopefully, finds him- or herself alive and well and still in possession of all his or her faculties. More suggests that you could also make back-up recordings of your brain states so that if you were run over by a bus and killed, you could call up your brain state from two weeks ago and reboot yourself minus the two weeks preceding the accident.

None of these schemes is even remotely practical with current technology, yet none is strictly impossible either, given our current understanding of the brain. As we have seen throughout this book, the gradual shift from biology to technology is already taking place. Retinal implants allow the blind to regain some sight; cochlear implants allow the deaf to hear; chips in the spinal cord and limbs restore basic sensory-motor function. It's easy to imagine a time when this kind of technological enhancement might extend to the brain itself; indeed, as we saw above, working junctions between brain cells and computer chips have already been made. Who can say where it will end?

But are these intimations of immortality real? While crude neuron–silicon interfaces already exist, it's far from certain that this kind of brain–computer hybrid would necessarily confer immortality. No one has the faintest idea of how consciousness emerges in biological brains, so we're unlikely to be able to transfer it into artificial ones. And if memory is contained in constantly shifting neural patterns, then transferring just a single snapshot from this continually changing tapestry is likely to be as effective as transferring memory proteins among worms: in short, not very.

A major drawback to the effectiveness of an artificial brain has to do with the fact that the biological brain is not regulated by electricity alone but by hormones and neurotransmitters as well; a lot of what goes on inside our heads is choreographed by the chemicals transmitted and absorbed by the synapses. Any non-biological brain aimed at sustaining a person's consciousness and identity would somehow need to model these biochemical processes as well.

The silicon soul described by Leary is actually not a captured consciousness at all, but an elaborate, interactive home video of history. Useful and entertaining as that might be, it's a far cry from life after death; and just how riveting would such a well-documented life be? Home videos are already boring enough. Imagine if the camera was recording *all the time*!

In any case, comprehensiveness is not the most important goal when it comes to achieving immortality, or superlongevity. Key moments and peak experiences make an individual unique and a life worth living and remembering, not an exhaustive catalogue of everything that person ever said or did. It's those few things they said and did that made a difference that really matter. As it turns out, we already have some pretty good technology that very effectively records these decisive instants: they're called books. How big an improvement would an immortal silicon soul be over, say, the novels of James Joyce or the poems of John Donne?

In his book *Dancing on the Grave*, British anthropologist Nigel Barley describes a trip to Africa during which a group of tribal elders explains their idea of reincarnation by visiting a local brewery. 'You could see returned bottles through a plate glass window,' Barley writes, 'entering via one door, whirling from machine to machine ... being endlessly refilled with squirting beer, relabelled and pushed out through another door ... "Life, death, spirit and body. Now you have seen," [the elders] said.'

Though it comes from Africa, this little episode is a fitting metaphor for Western attitudes towards death: the body is a machine and death is a spanner in the works; but you don't have to postulate the existence of an eternal soul to acknowledge that there may be some things in the universe, such as death, that we just can't fix. Let us not be too proud that our technological prowess may one day bring immortality. Never send to know for whom the computer crashes; it crashes for thee.

IN THE REALM OF THE SENSES

This chapter began with claims about a new genesis taking shape. It ends with a passage from the Book of Genesis: Let man 'have dominion over the fish of the sea, and over the fowl of the air, and over every living thing that moves upon the earth'.

In a performance first shown in 2000, multimedia artist Eduardo Kac translated this quotation from the Old Testament into Morse code and, by means of a conversion principle he devised himself, translated the dots and dashes of the Morse code into a strand of DNA he called 'the Genesis gene'. He used the four chemical letters of the DNA code, replacing the dots of Morse with the base cytosine, the dashes with the base thymine, the spaces between letters with guanine and the space between words with adenine. The Genesis gene was incorporated into the genomes of a colony of *E. coli* bacteria, where it made the bacteria glow when exposed to ultraviolet light. The ultraviolet light caused mutations in the bacteria, which altered the DNA in the Genesis gene, which in turn rearranged the letters in the quotation from the Book of Genesis.

As part of the performance, participants accessed the gallery via a webcam and were able to activate the ultraviolet light above a Petri dish containing the genetically modified bacteria. A video projection displayed a view of the bacterial division and interaction as seen through a microscope. At the end of the show the altered biblical sentence was decoded and translated back into English. Parts of it read: 'Let aan have dominion over the fish of the sea and over the fowl of the air and over every living thing that ioves ua eon the earth.'

Kac says,

> This passage from Genesis was chosen for what it implies about the dubious notion of divinely sanctioned human supremacy over nature. Morse code was chosen because, as first employed in radiotelegraphy, it represents the dawn of the information age, the genesis of global communications. The work confronts us with a dilemma: we have to take responsibility for changing the genetic structure of an organism, and for changing the word of God in the body of the bacteria.

This book has chronicled the ways in which people are changing the human body through technology, and endowing electronic devices with senses that might one day give them a soul. Some might consider this to be tampering with the work of God or interfering with nature in inappropriate and possibly dangerous ways. There are, of course, dangers in this kind of technology, just as there are in any new technology. It's commonplace to observe that technology is morally neutral; it's people who are good or evil and who put technology to good or evil uses. Morally neutral though it may be, technology does induce in people a desire to use it. Consider the incredible popularity of the mobile phone: no one felt the compulsion to communicate so much or so often – or so unnecessarily – until the means were available to do so. Technology creates demand as well as feeds it.

What new demands will this technology create? It's difficult to say, since many of the devices and interfaces are still so immature. It could well be decades before the most promising applications emerge. Nevertheless, entertainment is already sure to be one of the main sources of demand.

Joysticks and steering wheels fitted out with haptic feedback already exist. Imagine the possibilities when the emotional and physical effects of a video game could be induced through a direct interface with the brain, a future envisioned in the 1999 film *Existenz*. Gone will be the need for any apparatus at all; all the gameplayer will need is the proper electrical stimulation in the brain; and why stop at games when any type of novel sensation would be at your fingertips. Users could orchestrate their own brain states, calling up everything from opium-like highs to day-long orgasms at the press of a button.

More sinister applications are not hard to imagine either. The opportunities for covert surveillance and unwelcome monitoring will increase when the walls, and many other everyday objects, have ears, eyes, noses and mouths. Privacy will become even more fragile, and even more fiercely guarded, as a result; and when direct brain interface technology becomes more sophisticated, the threat of mind control will become real. Brainwashing will no longer require years of rigorous indoctrination, but some relatively straightforward neurosurgery and the appropriate computer program.

Yet as the Borg are fond of saying in *Star Trek: The Next Generation*, 'resistance is futile' – and just plain stupid. Technology will continue to advance into regions where no one has gone before, and this is as it should be. It's far wiser to engage in that process from the start to help shape its outcome than to recoil in horror from its possible consequences.

There is, in fact, nothing unnatural about technology. It has always been a part of human nature, since the first *Homo sapiens* used a stone to crack a nut and crush a skull. What's changing is that technology is moving from being something outside the body to something inside the body, from being a way we shape the external world to a way we shape the internal world of our perceptions, feelings and thoughts. Technology is becoming a part of our bodies rather than a mere extension of them. 'The relationship between technology and the body is frightening because the skin has always formed a barrier between inside and outside,' Kac says. 'Now that barrier is being crossed.'

This should not alarm us since nobody – and no body – is now or ever has been really secluded from the external environment. While the skin may be a barrier between the outside world and our internal organs, the senses themselves are portals designed to let the outside world in rather than keep it out. The senses enable us to perceive the world, but they also restrict our view. We can't see light in the ultraviolet part of the spectrum, for example, and we can't hear extremely low- or extremely high-frequency sounds; but other animals can. There's nothing unnatural about this, and we wouldn't be transgressing any laws of God or nature if we were to endow ourselves with these abilities as well. Birds do it. Bees do it. Why shouldn't we do it?

The senses themselves do not passively receive information about the world; they actively shape, distort and, in many ways, create what we perceive. When we see an object, for example, the brain actively composes the image, extending lines and rounding out corners where these aspects of the object itself are incomplete or indistinct. The brain makes its own reality based on the ambiguous, ill-defined and often contradictory information pouring in through the senses. As a result, sensory perceptions are modifications, impressions and recreations of what's 'out there' in the physical world. Moreover, human beings have been modifying their bodies for centuries, through cosmetic surgery, diet regimens, drug regimes (both recreational and medicinal) and exercise. Now this process of modification is being internalised.

If it's natural for our senses to actively create the world we perceive, why shouldn't we actively recreate our senses to enhance or change the way we experience that world? 'We have always asked what machines can do for us,' Kac says. 'Now might be the right time to ask what we can do together.'

Kac's work provides a glimpse of what man and machine can do together. In the 1999 work A–positive, Kac hooked himself up via an intravenous line to a machine. The work gives a contemporary twist to the artificial heart–lung machine, a device that maintained circulation during heart surgery by pumping the patient's blood away from the heart, oxygenating it and then returning it to the body. The artificial heart–lung was first successfully employed on a human being in 1953, and was one of the first true bionic devices. In A–positive, an intravenous line ferried Kac's blood to the machine, which extracted enough oxygen from it to support a small flame. In return, the machine deposited dextrose, a nutrient, into Kac's bloodstream and returned it to his body. Kac gave the machine enough oxygen to keep the flame alive, and the machine gave Kac enough energy in the form of dextrose to make up for his loss of blood.

That little tongue of flame is a fitting metaphor for how technology and biology are coming together to pry open a little further the doors of our perception. Norbert Wiener once advised, 'Render unto man the things which are man's and unto the computer the things which are the computer's.' That's sage counsel, indeed; but as this anatomy of the new bionic senses has shown, it's becoming increasingly difficult to tell the difference.

However our perceptions may be extended, enhanced or repaired in the future, we will still have the humble neuron to thank for keeping our senses burning bright. So let us sing the body electric, mindful of its perils as well as its possibilities, and lift up our voices in praise of the Holy Sprit.

NOTES AND REFERENCES

CHAPTER 1. THE SILICON SENSORIUM: AN INTRODUCTION

Page

1 *'It can also be maintained that it is best to provide the machine . . .'*. Turing, A. M., 'Computing Machinery and Intelligence', *MIND* (the Journal of the Mind Association), vol. LIX, no. 236, 1950, pp. 433–60.

4 *'Cybernetics'*. Wiener, Norbert, *The Human Use of Human Beings: Cybernetics and Society*, Da Capo Press, Inc., New York, 1954, p. 16.

4 *'Cyborg'*. Clynes, M., and Kline, N. 'Cyborgs and Space', in: *The Cyborg Handbook* (C. G. Hables, ed.), Routledge, London, 1995, p. 29.

7 *'With the arrival of electric technology . . .'*. McLuhan, Marshall, *Understanding Media: The Extensions of Man*, MIT Press, Cambridge, Mass., 1997, p. 219.

7 *'Any extension, whether of skin, hand, or foot . . .'*. Ibid., p. 4.

7 *'The framework itself changes . . .'*. Ibid., p. 43.

7 *'The human body becomes electric'*. Whitman, Walt, *Complete Poetry and Selected Prose*, Houghton Mifflin Company, Boston, 1959, p. 70.

CHAPTER 2. SIGHT: THE VISION THING

Page

12 *The MIVIP visual prosthetic*. Veraart, C., *et al.*, 'Visual sensations produced by optic nerve stimulation using an implanted self-sizing spiral cuff electrode', *Brain Research*, 813, 1998, pp. 181–6.

12 *The MIVIP and the optic nerve*. A visual prosthesis based on electrical stimulation of the optic nerve. Wanet-Defalque, M.-C., *et al.*, *Proceedings of the 5th Annual Conference of the International Functional Electrical Stimulation Society and 6th Triennial Conference 'Neural Prostheses: Motor systems'*. (T. Sinkjaer, D. Popovic and J. J. Struijk, eds.), 18–24 June 2000, pp. 146–8.

14 *Technological approaches to blindness*. One alternative to invasive approaches such as implants is sensory substitution, a technique that uses an intact sensory system to process information intended for the defective system. In addition to his work on the MIVIP, Claude Veraart also made a sensory substitution system to replace missing visual information with auditory information. Veraart and his team have developed an experimental prototype that consists of headphones and a miniature

head-mounted video camera connected to a PC. The video camera captures live images and the computer assigns each pixel of that image a specific tone, which it then compiles into an auditory translation of the visual scene and sends to the person through the headphones.

Subjects use the system in the same way as the MIVIP device. To explore the environment, patients must move their head from side to side and as the video camera sweeps the scene, the images it picks up are translated into sounds. A series of rapid tones heard while moving the head from left to right might indicate a wall ahead, while the same series heard while moving the head straight up and down might indicate a tree. Veraart and his team also established a set of parameters to help people using the system orient themselves. If the frequency of the sound is high, the object with which it is associated is in the upper half of the visual field; if the frequency is low, then it's in the bottom half. There are also different tones for objects on the left and on the right.

Ideally, sensory substitution systems will give blind people the ability to recognise and localise visual patterns to help them navigate unfamiliar environments. But critics note that the technology is often a distraction rather than an aid and suggest that these complex devices are not much of an improvement over the low-tech cane, a sensory substitution system that provides audible information by the way the sound of tapping changes in response to different objects. For more information, see: Arno, P., Capelle, C., Wanet-Defalque, M-C., Catalan-Ahumada, M., and Veraart, C., 'Auditory coding of visual patterns for the blind', *Perception*, 1999, volume 28, pp. 1013–29; and Capelle, C., Trullemans, C., Arno, P., and Veraart, C., 'A real-time experimental prototype for enhancement of vision rehabilitation using auditory substitution', *IEEE Transactions on Biomedical Engineering*, vol. 45, no. 10, October, 1998, pp. 1279–93.

15 *The best place for a visual prosthesis.* Rizzo, J. F., and Wyatt, J. L., 'Retinal Prosthesis', in: *Age-Related Macular Degeneration* (J. W. Berger, S. L. Fine and M. G. Maguire, eds.), Mosby, St Louis, 1998, pp. 413–32.

15 *The Harvard/MIT retinal implant.* Wyatt, J. L., and Rizzo, J. F., 'Ocular Implants for the Blind', *IEEE Spectrum*, vol. 33, May 1996, pp. 47–53.

Researchers at the University of Bonn in Germany have taken an approach similar to that of Rizzo and Wyatt. The Bonn retinal implant works like the Harvard/MIT device, but adds a retinal encoder located outside the body. The retinal encoder replaces the information-processing capability of the retina with a neural network. The retinal encoder transforms a visual scene into electrical signals, just like a biological retina, but the person with the implant can alter its function to produce the best possible perception. The subject looks at phosphene patterns – a circle, for example – and can change the parameters of the device until the image duplicates what they are told they should be seeing. The process is similar to tuning a television station to get the best reception. This feedback is stored on the computer's neural network so that the next time that phosphene pattern is encountered the image is picture perfect.

16 *'The technology has great potential . . .'.* Rizzo, J. F., and Wyatt, J. L., 'Prospects for a Visual Prosthesis', *Neuroscientist*, vol. 3, July 1997, pp. 251–62.

17 *The ASR.* Peachey, N. S., and Chow, A. Y., 'Subretinal implantation of semiconductor-based photodiodes: Progress and challenges', *Journal of Rehabilitation Research and Development*, vol. 36, no. 4, October 1999, pp. 371–6.

18 *Can subretinal implants be tolerated in the human eye?* Chow, A. Y., and Chow, V. Y., 'Subretinal electrical stimulation of the rabbit retina', *Neuroscience Letters* 225 (1997) pp. 13–16. See also: Peyman, G., *et al.*, 'Semiconductor Microphotodiode Array', *Ophthalmic Surgery and Lasers*, March 1998, vol. 29, no. 3, pp. 234–41.

19 *'The Dobelle eye'.* Dobelle, A., 'Artificial vision for the blind by connecting a television camera to the visual cortex', *Journal of the American Society of Artificial Internal Organs*, 2000; 46: 3–9.

21 *Alfred Smee on artificial eyes.* Cited in Dyson, George, *Darwin Among The Machines*, Penguin, London, 1997, p. 48.

23 *'Neuromorphic chips'.* Hahnloser, R. H. R., Sarpeshkar, R., Mahowald, M. A., Douglas, R. J., and Seung, H. S., 'Digital selection and analogue amplification co-exist in a cortex-inspired silicon circuit', *Nature*, vol. 405, 22 June 2000, pp. 947–51.

25 *How the VRD works.* Furness, T., European Community/National Science Foundation Position Paper: 'Toward Tightly-Coupled Human Interfaces', presented at First EC/NSF Advanced Research Workshop, 1–4 June 1999, Château de Bonas, Toulouse, France.

30 *'I have melded technology with my person ...'.* Mann, Steve, 'Cyborg Seeks Community', *Technology Review*, May–June 1999, pp. 36–42.

31 *'The WearComp'.* Mann, Steve, 'An historical account of the WearComp and WearCam inventions developed for applications in Personal Imaging', *IEEE Proceedings of the First ISWC*, October 1997, Cambridge, Massachusetts, pp. 66–73.

31 *'The VibraVest'.* Mann, Steve, 'VibraVest/ThinkTank: Existential Technology of Synthetic Synaesthesia for the Visually Challenged', *The Eighth International Symposium on Electronic Arts*, Art Institute of Chicago, September 1997. Available at *http://widow.artic.edu/webspaces/isea97/*.

31 *'Synthetic synaesthesia'.* Mann, Steve, 'Synthetic Synesthesia of the 6th and 7th Senses', *Proc. IEEE*, vol. 86, no. 11. Available at *http://hi.eecg.toronto.edu/hi.htm*.

33 *Homographic modelling.* Mann, Steve. Available at *http://wearcam.org/comparam.htm*.

35 *The Salk Institute lie detector.* 'Face the Music: How New Computers Catch Deceit', *Time*, 13 March 2000, pp. 58–9.

37 *'Microexpressions'.* When it comes to lie detecting, a person's tone of voice can be just as revealing as the expression on his face. A low tone of voice can suggest that a person is lying or stressed, while a higher pitch can mean excitement. Israeli inventor Amir Liberman has come up with a device – the Verdicator – that he claims enables anyone with a personal computer and a phone or microphone to catch a liar. The Verdicator delivers its results by analysing voice fluctuations that are usually inaudible to the human ear. When a person is under stress, anxiety may cause muscle tension and reduce blood flow to the vocal cords, producing a distinctive pattern of sound waves. Liberman has catalogued these patterns and programmed the Verdicator to distinguish between tones that indicate excitement, cognitive stress – the difference between what you think and what you say – and outright deceit. Once linked to a communications device and computer, the Verdicator monitors the subtle vocal tremors of your conversational partner and displays an assessment of that person's veracity on the screen.

CHAPTER 3. HEARING: THE SOUNDS OF SCIENCE

Page

41 *'Baldi teaches students to produce more accurate utterances'*. Cole, J. T., *et al.*, 'Tools for research and education in speech science', in: *Proceedings of the 14th International Congress of Phonetic Sciences*, San Francisco, August 1999.

42 *When hair cells die.* Unlike human beings, birds apparently do not become permanently deaf when their hair cells are destroyed. In a paper published in the *Proceedings of the National Academy of Sciences*, scientists from the University of Washington reported how support cells near the base of avian hair cells develop into new hair cells when the old ones die. The researchers are trying to understand how this mechanism works in the hope that the process could be recreated in humans. The studies might also shed light on the process of hair cell death itself. If this process could be interrupted, or reversed, some forms of deafness might be preventable.

Tinnitus, a ringing, whistling or hissing noise in the head that has no external cause, is another malady associated with damaged hair cells. The severity of the condition – the mechanisms of tinnitus are still so obscure that researchers are not even sure it's a disease – varies: some forms are as loud as jackhammers, while others sound more like rushing water or a swarm of bees. But everyone who has experienced tinnitus agrees that, in the most serious cases, it can be a maddening and psychologically devastating affliction. Statistics compiled by Tinnitus Action, a London-based organisation devoted to increasing public understanding of the condition, show that roughly one person in ten is affected by tinnitus. For about 15 per cent of sufferers, the noises inside their heads are disturbing enough to shatter their peace of mind, leading to everything from forgetfulness to severe depression and panic attacks.

Scientists don't really know why or how tinnitus happens. The condition is generally regarded as a symptom of something else, possibly a problem of the inner ear and its connection through the auditory pathway to the hearing centres of the brain. One potential trouble spot is the hair cells on the cochlea, which can be damaged by sustained exposure to loud noises. Indeed, one common form of tinnitus is the temporary and relatively mild ringing in the ears experienced after a rave or a rock concert. Other researchers suggest tinnitus is an amplified version of the brain's own background noise, a random crackling and sputtering of neurons that usually goes unnoticed. According to this theory, the neural static somehow slips through the filters in the brain that normally screen out irrelevant noise. As a result, the person is painfully aware of a sound that isn't really there.

Most tinnitus sufferers are eventually able to adjust to their condition. Physicians advise against unproven and potentially harmful treatments such as cortisone injections, drug regimes or hyperbaric oxygen therapy, in which patients sit in a pressurised chamber for 30 minutes breathing pure oxygen. Psychological warfare seems to be the most effective way to drown out the unwanted noise. Because tinnitus may be stress-related, some doctors recommend relaxation techniques such as yoga and meditation with the use of a device like a hearing aid, which produces white noise to mask the tinnitus.

44 *Sarpeshkar's silicon cochlea.* Sarpeshkar, R., 'Traveling Waves Versus Bandpass Filters: The Silicon and Biological Cochlea', *Symposium on Recent Developments in Auditory Mechanics*, Sendai, Japan, July 1999, pp. 216–22.

44 *The silicon cochlea.* In contrast to cochlear implants, the Personal Active Radio/Audio Terminal made by researchers in Trondheim, Norway, allows people to screen out unwanted noise. The device is an earpiece equipped with a microphone and miniature loudspeaker. The microphone can be programmed to pick up only selected sounds, which are then transmitted to the loudspeaker, while all other noises are shut out. For more information, see 'Tune in, tune out', *New Scientist*, 17 February 2001, p. 21.

46 *'Muu are ... different'.* 'Muu: Artificial Creatures as an Embodied Interface', Okada, M., Sakamoto, S., and Suzuki, N., *SIGGRAPH 2000*, Conference Abstracts and Applications, Computer Graphics Annual Conference Series, ACM SIGGRAPH, p. 91.

46 *'People like a good natter'.* For more on Okada's thoughts regarding the uses of gossip, see 'Talking Eye: Autonomous creatures for augmented chatting', Suzuki, N., Takeuchi, Y., Ishii, K., and Okada, M., *Robotics and Autonomous Systems*, 31 (2000), pp. 171–84.

50 *Prosody and speech recognition in the brain.* Steinhauer, K., Alter, K., and Friederici, A. D. (1999), 'Brain potentials indicate immediate use of prosodic cues in natural speech processing', *Nature Neuroscience*, 2 (2), pp. 191–6.

53 *'In the 1951 sci-fi classic* The Day the Earth Stood Still ...'. The answer is: 'Gort! Klaatu berada nikto.'

56 *'A ... portable computer ... is programmed with the universals of human mentality ...'.* Moravec, Hans, *Mind Children: The Future of Robot and Human Intelligence*, Harvard University Press, Cambridge, Mass., 1988, pp. 110–11.

57 *'Excuse me, but are you human?'.* Though hardly human, Neuro Baby is intended to mimic some of the behaviours of an infant. Swaddled in a blanket and staring up from a crib with its big blue eyes, Neuro Baby passes the time by whistling a little nursery rhyme. The tune is interrupted by an occasional yawn or hiccup. When someone comes along and speaks to it in a gentle, soothing voice, it smiles and giggles; if addressed in a low, angry tone, its face collapses into a frown and it starts to cry.

According to its creator, Naoko Tosa of the ATR Media Integration & Communications Research Laboratories near Kyoto, Japan, Neuro Baby is 'an automatic facial expression synthesiser that responds to expressions of feeling in the human voice. Neuro Baby lives inside a computer and communicates with others through its responses to inflections in human voice patterns.'

Neuro Baby responds to Japanese and English phrases and looks like a cross between the Michelin Man and a circus clown, with the cartoon face of a child, a pudgy body and a billowy white suit with big red buttons. When it's surprised, its big blue eyes bulge and its hands shoot up, palms forward, in an expression of helpless astonishment; when it's angry, its brow furrows and its face turns reddish purple.

Neuro Baby is able to simulate human emotional reactions because its artificial intelligence program has learned to associate specific tones of voice with the matching emotional states. The program knows, for example, that high-pitched cooing sounds usually correspond to affection and playfulness. So when Neuro Baby hears this, it infers that the appropriate response is a smile. Similarly, when loud, low tones occur Neuro Baby knows these are usually associated with anger so tears may be in order. Neuro Baby's facial expressions are generated by a computer animation programme that matches vocal inputs with the appropriate response. The digital infant can even make eye contact thanks to an eye-sensing system that tracks the position of a speaker's face.

For more on Neuro Baby and Tosa's other work, see Tosa, Naoko, 'Expression of emotion, unconsciousness with art and technology', in: *Affective Minds* (G. Hatano, N. Okada and H. Tanabe, eds.), Elsevier, Amsterdam, 2000, pp. 183–202.

58 *The sound of Von Kempelen's machine*. A brief history of speaking machines, by Hartmut Traunmüller of Stockholm University, can be found in 'Wolfgang von Kempelen's and the subsequent speaking machines' at *http://www.ling.su.se/staff/hartmut/kemplne.htm*.

59 *'The Voder . . . at the 1939 World's Fair'*. 'Now A Machine That Talks With The Voice of Man', *Science News Letter: The Weekly Summary of Current Science*, 14 January 1939, p. 19.

60 *'The Babel fish'*. Adams, Douglas, *The Hitchhiker's Guide to the Galaxy*, Pan Books, London, 1980, p. 50.

62 *Endangered languages*. For more on endangered languages, see Wurm, Stephen (ed.), *Atlas of the World's Languages in Danger of Disappearing*, UNESCO Publishing/Pacific Linguistics, Paris/Canberra, 1996, and Geary, James, 'Speaking in Tongues', *Time*, 7 July 1997, pp. 52–8. For more on language death, see Crystal, David, *Language Death*, Cambridge University Press, Cambridge, 2000.

CHAPTER 4. SMELL: ADVENTURES IN ODOUR SPACE
Page

66 *'The invention of odours . . .'*. De Montaigne, Michel. *The Complete Essays*, Translated by M. A. Screech, Penguin Books, London, 1991, pp. 353–4.

68 *'Smell-O-Vision'*. Dale Air Deodorising, a company based in Lytham, England, manufactures a device that is very much like Laube's malodorous smell brain. Known as the Ultra-Vortex, it too contains an array of fragrances or deodorising liquids that are introduced into an environment – a restaurant, shop or museum – as a very fine spray, usually through the building's air conditioning system. Dale Air not only manufactures the delivery system, it makes the scents as well. Visit the gift shop at Eden Camp, a Word War II museum in Malton, near York, and you can purchase vials of nostalgic odours such as 'the aroma of a U-Boat at sea', 'the aroma of a YMCA tea wagon' and even 'the aroma of a street at war' – all made by Dale Air.

68 *'Scent of Mystery'*. 'Nose Opera', *Time*, 29 February 1960, p. 98.

69 *'The scent of these armpits aroma finer than prayer'*. Whitman, Walt, *Complete Poetry and Selected Prose*, Houghton Mifflin Company, Boston, 1959, p. 42.

Whitman is biologically as well as poetically correct in pointing out that sweat can smell pleasant, or at least neutral. Fresh perspiration is, in fact, odourless. Only after skin bacteria start slurping up the sweat – and excreting waste – is body odour produced; and body odour is often far more efficacious than prayer, if not always as sweet. It has been shown, for example, that pheromones – odourless molecules that can influence emotions and behaviour – extracted from human perspiration can affect the regularity of a woman's menstrual cycle.

70 *Smell and spirituality*. For more on odours and spirituality, see 'Olfaction and Transition', in Howes, David (ed.), *The Varieties of Sensory Experience: A Sourcebook*

in the Anthropology of the Senses, University of Toronto Press, Toronto, 1991, pp. 128–47.

Howes is part of the Concordia Sensoria Research Team (CONSERT) at Concordia University in Montreal. CONSERT (*http://alcor.concordia.ca/~senses/*) is made up of a group of anthropologists who are investigating the senses across cultures and classes as a way to bring to the fore previously suppressed aspects of aesthetic experience. They intend their work to constitute the first comprehensive study of the role of smell, taste and touch in art.

70 *Proust on taste and smell*. Proust, Marcel, *Remembrance of Things Past*, vol. I, translated by C. K. Scott Moncrieff and Terence Kilmartin, Random House, New York, 1981, pp. 50–51.

70 *'The rhinencephalon'*. In his treatise *The Olfactory Organ* Galen was the first to conclude that the brain and not the nose is the actual organ of smell. Other creatures use other appendages as conduits for olfaction. Crabs, for example, have odour receptors in their claws, while the silkworm moth *Bombyx mori* sniffs with its antennae.

70 *Foetal smell*. Schaal, B., Marlier L., and Soussignan, R., 'Earlier-Than-Early Odour Learning: Human Newborns Remember Odours From Their Pregnant Mother's Diet', in: *13th International Symposium Olfaction & Taste and 14th European Chemoreception Research Organisation Congress* (Persuad, K. C., and Van Toller, S., eds.), ECRO, 2000, pp. 68–9.

71 *'If you ask someone to put a name to a smell . . .'*. Vroon, Piet, *Smell: The Secret Seducer*, Farrar, Straus and Giroux, New York, 1997, p. 108.

Not all vocabularies are equally impoverished when it comes to olfaction. Quechua, the ancient Incan language that is still spoken in many parts of the Andes, has more terms dealing with smell than most Western tongues. The word *mutquini*, for example, means 'to smell something'; *muccani* means 'to emit an odour'; *cuntuni* means 'to smell good'; *curayani* means 'to smell bad' and *aznapayani* means 'to give a bad odour to another'. The Batek Negrito tribe of peninsular Malaysia classify virtually everything in their environment by smell, including the sun (which reeks of raw meat) and the moon (which smells like flowers). In the Eastern Arrernte language of Central Australia the simple, sensual word *nyimpe* denotes 'the smell of rain'.

For more on Quechua verbs 'to smell', see Howes, David, and Classen, Constance, 'Healing Scents of the Andes', *The International Journal of Aromatherapy*, Winter 1993, vol. 5, no. 4, pp. 19–23. For more on Batek Negrito osmology, see Howes, David (ed.), *The Varieties of Sensory Experience: A Sourcebook in the Anthropology of the Senses*, p. 283.

72 *Linda Buck and the genetics of smell*. Malnic, B., Hirono, J., Sato, T., and Buck, L. B., 'Combinatorial receptor codes for odour', *Cell*, 1999, 96, pp. 713–23.

The Human Olfactory Receptor Data Exploratorium (HORDE) contains information on the known olfactory receptor proteins, their structure, function and evolution as well as a set of analysis tools (*http://bioinformatics.weizmann.ac.il/HORDE/*).

73 *'There, on the surface of the cilium or cell wall . . .'*. Lettvin, J. Y., and Gesteland, R. C., 'Speculations on Smell', in: *Cold Spring Harbor Symposia on Quantitative Biology, Volume XXX, Sensory Receptors*, Cold Spring Harbor, 1965, p. 223.

73 *The olfactory system of the northern grass frog*. Kleene, S. J., and Gesteland,

R. C., 'Calcium-activated chloride conductance in frog olfactory cilia', *Journal of Neuroscience*, 1991, 11, pp. 624–9.

73 *'Almost every odour seems to affect almost every receptor ...'*. Lettvin, J. Y., and Gesteland, R. C., 'Speculations on Smell', pp. 219 and 224.

75 *Schizophrenia breath test.* 'Breath Test', *New Scientist*, 5 August 2000, p. 18.

76 *Asthma breath detection.* Kharitonov, Sergei A., and Barnes, Peter J., 'Exhaled Breath Analysis', in: *New Drugs for Asthma, Allergy and COPD. Progress in Respiratory Research*, vol. 31 (Hansel, T. T., and Barnes, P. J., eds.), Basel, Karger, 2000.

77 *The Kauer artificial nose.* Dickinson, T. A., White, J., Kauer, J. S., and Walt, D. R., 'A Chemical-detecting system based on a cross-reactive optical sensor array', *Nature*, 1996, 382, pp. 697–700.

77 *Odour detection and neural networks.* Johnson, S. R., Sutter, J. M., Engelhardt, H. L., Jurs, P. C., White, J., Kauer, J. S., Dickinson, T. A., and Walt, D. R., 'Identification of multiple analytes using an optical sensor array and pattern recognition neural networks', *Analytical Chemistry*, 1997, 69, pp. 4641–8.

78 *Detecting hazardous substances.* Not all artificial noses are electronic. Jan van der Pers of Syntech/VDP Laboratories in the Netherlands has invented an entomological nose. In a strange variation on the canary in a coal mine, he conditions insects to perform as programmable odour sensors. The insects are trained to respond to a specific odour – carbon monoxide, for example – and radar is used to detect the telltale movements of their heads and antennae that indicate the odour has been encountered.

78 *'The zNose'.* Staples, E. J., 'The zNose™, A New Electronic Nose Using Acoustic Technology', Acoustical Society of America, December 2000, Paper Number 2aEA4.

80 *Smell over the Web.* Researchers at Michitaka Hirose's lab at the University of Tokyo have developed a head-mounted system that delivers odours right under your nose. The somewhat unwieldy device hooks around the ears like an enormous pair of glasses. Two loops of plastic tubing swoop under your nose like tusks. A steady stream of air gushes from the tubes, which terminate in fragrance-filled bottles at the other end. This olfactory helmet contains sensors that recognise certain scent-enabled objects in the lab environment. Look at a bottle of brandy on a shelf and that smell is transmitted through the tubes; look at a pile of oranges and that scent fills your nostrils. If the device could be made less cumbersome and, ideally, wireless, it might well liberate scent technology from the desktop.

83 *'Can you smell a ripe peach online?'.* Underhill, Paco, *Why We Buy: The Science of Shopping*, Orion Business Books, London, 1999, p. 217.

84 *Shopping 'involves experiencing ...'.* Ibid., p. 161.

CHAPTER 5. TASTE: FUN WITH ELECTRONIC TONGUES
Page

85 *The 'Futurist culinary revolution'.* Marinetti, Filippo Tommaso, *The Futurist Cookbook*, Trefoil Publications, London, 1989, p. 21.

85 *'Every person has the sensation of eating ... works of art'.* Ibid., p. 133.

85 *The 'economical dinner'.* Ibid., p. 114.

86 *A 'tactile dinner party'.* Ibid., pp. 125–6.

87 *'I am not only convinced ...'.* Brillat-Savarin, Jean-Anthelme, *The Physiology of Taste*, Penguin Classics, London, 1994, p. 41.

87 *'Anosmia'*. Anosmia is just one of a range of olfactory disorders that can wreak havoc on a person's sense of taste. Sylvia Jones, 74, who lives in Devon, England, fell in her garden ten years ago and, like Colin Berry, fractured her skull. Three days later she became aware of an awful stench in her nose and an awful taste in her mouth, both of which persist to this day.

Jones has aliosmia, the presence of an inappropriate bad smell, and phantogeusia, the presence of a phantom bad taste. Her condition is so severe that eating a meal is 'horrid beyond words', she says. 'I never eat any meat or spices. They always taste utterly foul.' The most she can manage to force down is some plain pasta, peas, perhaps a tiny slice of ham and some bread and butter. 'There is nothing in this world I put in my mouth that gives me pleasure,' she says. 'Even water tastes stagnant.'

Before the accident Jones's sense of smell was so acute that she was known as the 'family bloodhound'. Her mother and grandmother ran a florist's shop, so Jones had plenty of practice in identifying scents. She once discovered dry rot in her daughter's house by smell alone. Her family thought she was crazy, but she insisted that they take up a floorboard and, sure enough, there underneath was a little field of revolting fungi. 'What breaks my heart more than anything,' Jones says, 'is that I can no longer smell the wild honeysuckle when I go for a walk. My distorted smell and associated problems are like a bereavement that I find so hard to come to terms with.'

88 *'Coffee . . . contains more than 1000 different chemical components'*. Toko, Kiyoshi, 'A Taste Sensor', *Measurement Science & Technology*, 9, 1998, p. 1919.

88 *'About the size of a very large blender'*. Toko, Kiyoshi, 'Electronic Sensing of Tastes', in: *Sensors Update* (H. Baltes, W. Göpel and J. Hesse, eds.), vol. 3, 1996, pp. 131–60.

90 *Tasting beer with an electronic tongue*. Toko, Kiyoshi, 'Taste Sensor', *Transducers '99*, June 1999, pp. 58–61.

91 *Willie Wonka's Television Chocolate*. Dahl, Roald, *Charlie and the Chocolate Factory*, Puffin Books, London, 1995, pp. 159–61.

91 *'A more complete sensory experience'*. In addition to taste, TriSenx technology allows the sensations of smell and touch to be recreated over the Internet. Smells are brewed in much the same way as the Aerome ScentController, while tactile sensations are elicited by means of conductive polymer fibres that sprout tiny hair-like structures to stimulate the nerves in the fingertips. To get in touch with these feelings, users insert a hand through an opening in the TriSenx device, click on a touch-enabled website – an icon of a piece of fabric, for example, on a textile site – and then just sit back and let their fingers do the talking. The polymer fibres inside the box respond to the incoming signals to replicate the desired tactile sensation. Other textures are created by storing samples of different materials on a rotating spool, while heat and cold are generated through micro-heating and refrigeration units.

93 *'The Texas tongue'*. Savoy, S., *et al.*, 'Solution-based analysis of multiple analytes by a sensor array: Toward the development of an electronic tongue', SPIE Conference on *Chemical Microsensors and Applications*, SPIE, vol. 3539, Boston, November 1998.

95 *Nabokov's synaesthesia*. Nabokov, Vladimir, *Speak, Memory*, Perigree Books, New York, 1979, p. 35.

95 *Rimbaud's synaesthesia*. Rimbaud, Arthur, *Complete Works*, translated by Paul Schmidt, Harper & Row Publishers, New York, 1976, p. 123.

96 *Tesla's synaesthesia*. Tesla, Nikola, *My Inventions*, Barnes & Noble Books, New York, 1995, p. 36.

96 *Theories of synaesthesia*. Baron-Cohen, Simon, and Harrison, John E. (eds.), *Synaesthesia: Classic and Contemporary Readings*, Blackwell Publishers Ltd, Oxford, 1997, pp. 109–22.

97 *'The Dogon people of Mali'*. Classen, Constance, Howes, David, and Synnott, Anthony (eds.), *Aroma: The Cultural History of Smell*, Routledge, London, 1994, p. 119.

98 *The tongue tap*. Bradley, R. M., Xianghui, C., Tayfun, A., and Najafi, K., 'Long-term chronic recordings from peripheral sensory fibres using a sieve electrode array', *Journal of Neuroscience Methods*, 73 (1997), pp. 177–86.

CHAPTER 6. TOUCH: THE WORLD IS YOUR INTERFACE

Page

100 *'In the electric age . . .'*. McLuhan, Marshall, *Understanding Media: The Extensions of Man*, the MIT Press, Cambridge, Mass., and London, 1997, p. 47.

104 *'Functional electrical stimulation'*. Sinkjaer, T., 'Integrating Sensory Nerve Signals into Neural Prosthesis Devices', *Neuromodulation*, volume 3, number 1, 2000, pp. 32–8.

105 *Peripheral nerves and cuff electrodes*. Sinkjaer T., Haugland M., Struijk, J., and Riso, R., 'Long-term cuff electrode recordings from peripheral nerves in animals and humans', in: *Modern Techniques in Neuroscience* (U. Windhorst and H. Johansson, eds.), Springer-Verlag, 1999, chapter 29.

106 *Freehand effectiveness*. Taylor, P., Esnouf, J., and Hobby, J., 'Clinical Experience of the NeuroControl Freehand System', *Proceedings of the 5th Annual Conference of the International Functional Electrical Stimulation Society and 6th Triennial Conference 'Neural Prostheses Motor Systems'* (T Sinkjaer, D. Popovic and J. J. Struijk, eds.), 18–24 June 2000, pp. 88–91.

107 *'Sensitive prosthetic'*. Riso, R., 'Strategies for providing upper extremity amputees with tactile and hand position feedback – moving closer to the bionic arm', *Technology and Health Care* 7 (1999), IOS Press, pp. 401–409.

108 *Sensory feedback for prosthetics*. Inmann, A., and Haugland, M., 'A Flexible, Portable FES Hand Grasp Neuroprosthesis Incorporating Natural Sensory Feedback', *Proceedings of the 5th Annual Conference of the International Functional Electrical Stimulation Society and 6th Triennial Conference 'Neural Prostheses: Motor Systems'* (T. Sinkjaer, D. Popovic and J. J. Struijk, eds.), 18–24 June 2000, pp. 382–4.

108 *Prosthetic hand grasp control*. Santa-Cruz, M. C., Riso, R. R., and Sepulveda, F., 'Evaluation of Neural Network Parameters Towards Enhanced Recognition of Naturally Evoked EMG for Prosthetic Hand Grasp Control', *Proceedings of the 5th Annual Conference of the International Functional Electrical Stimulation Society and 6th Triennial Conference 'Neural Prostheses: Motor Systems'* (T. Sinkjaer, D. Popovic and J. J. Struijk, eds.), 18–24 June 2000, pp. 436–9.

110 *The parietal reach region*. Batista, A. P., *et al.*, 'Reach Plans in Eye-Centered Coordinates', *Science*, 9 July 1999, volume 285, pp. 257–60. See also, Snyder L. H., Batista, A. P., and Andersen, R. A., 'Coding of intention in the posterior parietal cortex', *Nature*, volume 386, 13 March 1997, pp. 167–70.

112 *Transmitting monkey brain signals over the Internet.* Nicolelis, Miguel A. L., *et al.*, 'Real-time prediction of hand trajectory by ensembles of cortical neurons in primates', *Nature*, vol. 408, 16 November 2000, pp. 361–5.

112 *Hybrid brain–machine interfaces.* Nicolelis, Miguel A. L., 'Actions from Thoughts', *Nature*, vol. 409, 18 January 2001, pp. 403–407.

112 *'Phantom limbs'.* Ramachandran, V. S., and Blakeslee, Sandra, *Phantoms in the Brain*, Fourth Estate, London, 1999, p. 22.

113 *Machines were to be 'regarded as a part of man's own physical nature . . .'.* Butler, Samuel, *Erewhon*, Penguin, London, 1985, p. 223.

119 *BIONs.* Loeb, G. E., and Richmond, F. J. R., 'BION Implants for Therapeutic and Functional Electrical Stimulation', in: *Neural Prostheses for Restoration of Sensor and Motor Function* (J. K. Chapin, K. A. Moxon and G. Gaal, eds.), CRC Press, Boca Raton, FL, 2000.

119 *BIONs and TES.* Richmond, F. J. R., Bagg, S. D., Olney, S. J., Dupont, A. C., Creasy, J., and Loeb, G. E., 'Clinical Trial of BIONs™ for Therapeutic Electrical Stimulation', *Proceedings of the 5th Annual Conference of the International Functional Electrical Stimulation Society*, Aalborg University, Denmark, 17–23 June 2000, pp. 95–7. See also: Dupont, A. C., Bagg, S. D., Creasy, J. L., Romano, C., Romano, D., Loeb, G. E., and Richmond, F. J. R., 'Clinical Trials of BION Injectable Neuromuscular Stimulators', *RESNA*, Reno, Nevada, 22–6 June 2001; and Dupont, A. C., Bagg, S. D., Creasy, J. L., Romano, C., Romano, D., Loeb, G. E., and Richmond, F. J. R., 'Clinical Trials of BION Injectable Neuromuscular Stimulators', *IFESS-01*, Cleveland, Ohio, 16–21 June 2001.

128 *'inTouch'.* Brave, S., Ishii, H., and Dahley, A., 'Tangible Interfaces for Remote Collaboration and Communication', in: *Proceedings of CSCW '98*, (Seattle, Washington, November 1998), ACM Press, pp. 169–78.

132 *'HapticGEAR'.* Hirose, M., Ogi, T., Yano, H., and Kakehi, N., 'Development of Wearable Force Display (HapticGEAR) for Immersive Projection Displays', *Virtual Reality '99 Conference* (L. Rosenblum, P. Astherimer and D. Teichmann, eds.), IEEE, 1999, p. 79.

133 *Adding sight and sound to haptics.* Salisbury, J. K., and Srinivasan, M. A., 'Phantom-Based Haptic Interaction with Virtual Objects', *IEEE Computer Graphics and Applications*, vol. 17, no. 5, 1997, pp. 6–10.

133 *'The Multimedia Virtual Laboratory'.* Hirose, M., Ogi, T., and Yamada, T., 'Integrating Live Video for Immersive Environments', *IEEE MultiMedia*, July–September 1999, pp. 14–22.

134 *Tactile displays.* Craig Chanter and Ian Summers of the University of Exeter in the UK have created a tactile display that generates a range of stimulation patterns on the fingertip by means of a hundred tiny pins, each powered by a tiny motor. The University of Exeter device consists of a one-square-centimetre array of the hundred contactor pins, on which the fingertip is placed. Each contactor pin is driven by a piezoelectric motor, and each motor can be individually programmed to provide a specific stimulus for a specific duration. Tactile sensations are produced by time-varying patterns of vibration delivered through the pins. Similar tactile sensations may be produced through electrical stimulation of the nerves that lead from the touch receptors in the skin.

A possible use of this technology is in a tactile hearing aid. This kind of device is a non-invasive and less expensive alternative to the cochlear implants we encountered in the chapter on 'Hearing', though current tactile aids can't begin to match the implants' performance. Their primary application is to support lip-reading, providing additional information on those parts of speech that can't be seen on the lips. An aid based on the Chanter and Summers technology might transmit more information than current tactile devices or could, for example, translate what a person is saying into Braille and deliver the Braille through the tactile display.

134 *Net sex.* 'Take your partners', *New Scientist*, 20 January 2001, p. 7.

Implanted electronics could soon make partners superfluous when it comes to sexual stimulation. A North Carolina surgeon stumbled across an orgasm-generating device while treating female patients for back pain. He discovered a spot on the spine that when stimulated with electrodes produced impulses that can trigger orgasm. To make an orgasm machine, a signal generator would have to be implanted under the skin while the electrical impulses could be controlled by a hand-held remote. The inventor hopes the device could be useful in treating sexual dysfunction. For more information, see 'Push my button', *New Scientist*, 10 February 2001, p. 23.

CHAPTER 7. MIND: THE SIXTH SENSE

Page

136 *'This is an idea with which I have toyed before . . .'.* Wiener, Norbert, *God & Golem, Inc.: A Comment on Certain Points where Cybernetics Impinge on Religion*, the MIT Press, Cambridge, Mass., 1990, p. 36.

136 *Patterns that persist.* Grand, Steve, *Creation: Life and How To Make It*, Weidenfeld & Nicolson, London, 2000, p. 6.

138 *How to find out if machines can have minds.* Edelman, Gerald M., *Bright Air, Brilliant Fire*, Penguin Books, London, 1992, pp. 189–90.

For more on Edelman's theories of consciousness, see Edelman, Gerald M., *Bright Air, Brilliant Fire*, Penguin Books, London, 1992, and Edelman, Gerald M., and Tononi, Giulio, *Consciousness: How Matter Becomes Imagination*, Allen Lane, the Penguin Press, London, 2000.

139 *Grand thinks they are alive.* This opinion is shared by members of the Equal Rights for Norns (ERFN) organisation, a group of people who believe that Norns deserve the same rights and respect as other living things. There is also a more shadowy group called the Norn-Torturers, a cadre of players led by the 'Anti-Norn' who abuse their Norns and post the results on the Web.

142 *The Norn neural network 'deserves to be called a brain'.* Grand, Steve, *Creation: Life and How To Make It*, p. 172.

142 *'Even though we feel in our heads like we are one person'.* Ibid., p. 93.

144 *'I do not believe these creatures are conscious . . .'.* Ibid., p. 212.

144 *'I believe that life can be created where there was none before'.* Ibid., p. 10.

145 *'Where consciousness is concerned, brains matter crucially'.* Searle, John R., *The Mystery of Consciousness*, Granta Books, London, 1997, p. 191.

150 *'I get kind of irritated with people who make a neuron . . .'.* Gray, C. H., 'An Interview with Jack E. Steele', in: *The Cyborg Handbook* (C. G. Hables, ed.), Routledge, London, 1995, p. 67.

150 *'The essence of consciousness is that it consists in internal qualitative, subjective mental processes'.* Searle, John R., *The Mystery of Consciousness*, p. 204.

150 *A 'parasite upon the machines . . .'.* Butler, Samuel, *Erehwon*, Penguin, London, 1985, p. 206.

150 *Tama the robotic cat.* Shibata, T., Tashima, T., and Tanie, K., 'Emergence of emotional behaviour through physical interaction between humans and artificial emotional creatures', in: *Affective Minds* (G. Hatano, N. Okadaand and H. Tanade, eds.), Elsevier, Amsterdam, 2000, pp. 232–9.

152 *Tama's artificial mind.* Ushida, H., Hirayama, Y., and Nakajima, H., 'Emotion Model for Life-like Agent and Its Evaluation', *Proceedings of the Fifteenth National Conference on Artificial Intelligence (AAAI-98)*, Madison, Wisconsin, 26–30 July 1998, pp. 62–9.

153 *WAMOEBA's endocrine system.* Ogata T., Matsuyama, Y., and Sugano, S., 'Acquisition of internal representation in robots – toward human–robot communication using primitive language', *Advanced Robotics*, vol. 14, no. 4, pp. 277–91 (2000). See also, Ogata, T., and Sugano, S., 'Emotional Communication Between Humans and Robots – Consideration of Primitive Language in Robots', in: *Proceedings of IEEE/RSJ International Conference on Intelligent Robots and Systems* (IROS '99), October 1999, pp. 870–75.

154 *WAMOEBA's internal states.* Sugano, S., and Ogata, T., 'Emergence of Mind in Robots for Human Interface – Research Methodology and Robot Model', in: *Proceedings of IEEE International Conference on Robotics and Automation* (ICRA '96), 1996, pp. 1191–8. See also, Ogata, T., and Sugano, S., 'Emotional Communication Robot: WAMOEBA-2R – Emotion Model and Evaluation Experiments', in: *Proceedings of IEEE/RAS International Conference on Humanoid Robots* (Humanoid 2000), September 2000, paper no. 93.

155 *'Intelligence cannot be separated from the subjective experience of a body . . .'.* Brooks, R. A., and Stein, L. A., 'Building Brains for Bodies', *Autonomous Robots*, vol. 1, no. 1, November 1994, pp. 7–25. Also published as *MIT AI Lab Memo 1439*, August 1993. See also, Brooks R. A., 'Intelligence Without Representation', *Artificial Intelligence Journal* (47), 1991, pp. 139–59.

155 *Kismet's drives and desires.* Breazeal, C., and Scassellati, B., 'How to build robots that make friends and influence people', *IROS99*, Kyonjiu, Korea, 1999. See also, Breazeal, C., 'Robot in Society: Friend or Appliance?' in: *Agents99 workshop on emotion-based agent architectures*, Seattle, WA, 1999, pp. 18–26.

158 *'In attempting to construct such [conscious] machines . . .'.* Turing, A. M., 'Computing Machinery and Intelligence', *MIND* (the Journal of the Mind Association), pp. 433–60.

158 *'For the idea that God's supposed creation of man and the animals . . .'.* Wiener, Norbert, *God & Golem, Inc.*, p. 47.

159 *Valentino Braitenberg's Vehicles.* Braitenberg, Valentino, *Vehicles: Experiments in Synthetic Psychology*, MIT Press, Cambridge, Mass., 1994.

160 *A working synapse between two snail neurons mounted on a chip.* 'Interfacing a silicon chip to pairs of snail neurons connected by electrical synapses', Jenkner, M., Müller, B., and Fromherz, P., *Biological Cybernetics*, 84 (2001), pp. 239–49.

161 *The controlled formation of snail synapses on a chip.* 'Electrical synapses by guided growth of cultured neurons from the snail *Lymnaea stagnalis*', Prinz, A. A., and Fromherz, P., *Biological Cybernetics*, 82, L1–L5 (2000).

161 *'These experiments show how large a gap really exists . . .'.* 'Neuron–Silicon Junction or Brain–Computer Junction?' Fromherz, Peter, *Ars Electronica 97*, pp. 158–61, 1997.

165 *Off-loading cognitive tasks.* Dennett, Daniel C., *Kinds of Minds: Towards An Understanding of Consciousness*, Phoenix, London, 1997, pp. 177–8.

165 *Saint Augustine on memory.* Augustine, *The Confessions*, translated by Maria Boulding, O. S. B., Hodder & Stoughton, London, 1997, pp. 246–7.

Interestingly, Saint Augustine includes in the chapter on memory a consideration of the five senses – and the temptations to which they lead. Augustine himself was a very sensual person, determined to refrain from 'concupiscence of the flesh and concupiscence of the eyes'. He liked good food, fine wine, music and visual beauty. For a fascinating and concise biography of Saint Augustine, see Wills, Garry, *Saint Augustine*, Weidenfeld & Nicolson, London, 1999.

166 *How memories are formed.* For more on memory, see Rose, Steven, *The Making of Memory: From Molecules to Mind*, Bantam Books, London, 1995, and Squire, Larry R., and Kandel, Eric R., *Memory: From Mind to Molecules*, Scientific American Library, New York, 1999.

166 *Long-term memories.* The déjà vu experience is one of the most perplexing anomalies of memory. It happens to us all at one time or another: that uncanny feeling of having experienced something before, but being unable to recall exactly when or where. The term déjà vu comes from the French for 'already seen', and is defined by psychiatrists as 'any subjectively inappropriate impression of familiarity of a present experience with an undefined past'. But that clinical description hardly does justice to the eerie sense of mystery and unease we feel during such an episode of inexplicable recognition.

In *The Psychopathology of Everyday Life*, Freud placed the déjà vu experience in the category of the miraculous and predicted that the subject would merit the most exhaustive treatment. But because it is an evanescent phenomenon, déjà vu is notoriously difficult to study. Most theories are speculative at best. Psychoanalysts, for example, maintain that déjà vu has to do with defence mechanisms and wish fulfilment. According to this theory, déjà vu is the expression of a wish to repeat a past experience – but this time with a more satisfactory outcome. Parapsychologists, on the other hand, suggest its a fleeting glimpse of some past life. But as the brain's mechanisms for learning and memory become better understood, scientists are proffering more plausible – though still preliminary – explanations of this strange and miraculous act of recollection.

A model for one such explanation is the hologram. In a hologram, a kind of three-dimensional photograph, each point in the image contains all the data necessary to reconstruct the image as a whole. 'Even the smallest fragment will give the complete picture,' says Herman Sno, a psychiatrist at De Heel Hospital outside Amsterdam who has made an extensive study of the scientific literature on déjà vu. 'But the smaller the fragment, the less sharp the picture will be.' If memories are indeed stored in the brain as a kind of hologram, each part of the memory contains all the sensory and emotional data needed to recall the entire original experience. A single detail – the sound of a child's voice, for example, or the smell of a lover's clothing – can evoke the complete remembered scene. According to this model, déjà vu occurs when a detail from a current experience so strongly resembles a detail from a previous experience that a full-blown memory of the past event is conjured up. 'As a result of the mismatching,' says Sno, 'the brain mistakes the present for the past. You feel certain you've seen the picture before.'

Another potential explanation involves a glitch in the exquisitely timed processes of perception and cognition. This theory proposes that sensory impressions of a current experience get detoured in the brain and are not immediately perceived. The information is, however, stored as a memory. 'This split-second delay in cognition creates the unsettling impression that the event is being experienced and recalled simultaneously,' says Sno. Whether it's a slippage of timing, a mental hologram or something else entirely, déjà vu will remain one of the mind's most tantalising and elusive tricks.

168 'Funes the Memorious'. Borges, Jorge Luis, Labyrinths: Selected Stories and Other Writings, New Directions, New York, 1964, pp. 59–66.

169 'Convergence zones'. Damasio's theories on memory and consciousness can be found in two books: Descartes' Error: Emotion, Reason and the Human Brain, Papermac, London, 1996; and The Feeling of What Happens: Body, Emotion and the Making of Consciousness, William Heinemann, London, 2000.

172 'Memory theatre'. Yates, Frances A., The Art of Memory, Pimlico, London, 1999.

172 'The memex'. All excerpts from Bush's article are from Bush, Vannevar, 'As We May Think', the Atlantic Monthly, July 1945, vol. 176, no. 1, pp. 101–108.

173 Computers, the brain and the World Wide Web. Berners-Lee, Tim, Weaving the Web: The Past, Present and Future of the World Wide Web by Its Inventor, Orion Business, London, 1999, pp. 3–4.

174 'Wholly new forms of encyclopedias . . .'. In The Muse in the Machine: Computerizing the Poetry of Human Thought (The Free Press, New York, 1994, p. 132), computer scientist David Gelernter proposes a device very similar to the memex: a computer whose 'principal goal is . . . to simulate memory sandwiching . . . In other words, the goal is to fetch memories in response to a probe, sandwich them together and peer through the whole bundle at once, notice the common features that "emerge strongly" in the overlay, and where it's appropriate, pick out interesting emergent details and probe further.' Like Bush, Gelernter sees these computers assisting in making medical diagnoses, forming legal opinions and carrying out financial analyses.

179 Hans Moravec on pattern-identity. Moravec, Hans, Mind Children: The Future of Robot and Human Intelligence, Harvard University Press, Cambridge, Mass., 1988, p. 117.

181 Let man 'have dominion over the fish of the sea . . .'. Genesis 1: 26.

185 'Render unto man the things which are man's . . .'. Wiener, Norbert, God & Golem, Inc.: A Comment on Certain Points where Cybernetics Impinge on Religion, p. 73.

FIGURE CREDITS

Figure 1. Copyright Claude Veraart.

Figure 2. Reprinted by permission from 'Prospects for a Visual Prosthesis', *Neuroscientist*, vol. 3, July 1997, pp. 251–62. Copyright (1997) Sage Publications, Inc.

Figure 3. Copyright Optobionics.

Figure 4. Copyright Advanced Bionics.

Figure 6. Reprinted by permission from Toko, Kiyoshi, 'Electronic Sensing of Tastes', *Sensors Update* (Baltes, H., Gåpel, W., and Hesse, J., eds.), vol. 3, 1996, pp. 131–60. Copyright (1996) John Wiley & Sons, Inc.

Figure 7. Reprinted by permission from Savoy, S., et al., 'Solution-based Analysis of Multiple Analytes by a Sensor Array: Toward the Development of an Electronic Tongue', *SPIE* Conference on Chemical Microsensors and Applications, *SPIE*, vol. 3, November 1998, p. 539. Copyright (1998) *SPIE*.

Figure 8. Event for Extended Body and Walking Machine, Cyborg Frictions, Dampfzentrale, Bern, 1999. Photo by Dominik Landwehr. Machine construction by F18, Hamburg.

Figure 9. Reprinted by permission from *Technology and Health Care*, vol. 7, No. 6, 1999, pp. 393–9. Copyright (1999) IOS Press.

Figure 10. Reprinted by permission from *Nature*, vol. 408, 2000, pp. 361–5. Copyright (2000) Macmillan Magazines Ltd.

Figure 11. Reprinted by permission from Loeb, G. E., *et al.*, 'BION system for distributed neural prosthetic interfaces', *Medical Engineering & Physics*, vol. 23, 2001, pp. 9–18 (figure 1). Copyright (2001) Elsevier Science.

Figure 12. Copyright Steve Grand.

Figure 13. Reprinted by permission from Vassanelli, S., Fromherz, P., 'Transistor Probes Local Potassium Conductances in the Adhesion Region of Cultured Rat Hippocampal Neurons', *Journal of Neuroscience*, vol. 19, 1999, pp. 6767–73. Copyright (1999) Society of Neuroscience.

BIBLIOGRAPHY

Ackerman, Diane, *A Natural History of the Senses*, Phoenix, London, 1996.

Adams, Douglas, *The Hitchhiker's Guide to the Galaxy*, Pan Books, London, 1980.

Aleksander, Igor, *Impossible Minds: My Neurons, My Consciousness*, Imperial College Press, London, 1996.

——, *How To Build A Mind*, Weidenfeld & Nicolson, London, 2000.

Augustine, *The Confessions*, translated by Maria Boulding, O.S.B., Hodder & Stoughton, London, 1997.

Barley, Nigel, *Dancing on the Grave: Encounters with Death*, Abacus, London, 1997.

Baron-Cohen, Simon, and Harrison, John E. (eds.), *Synaesthesia: Classic and Contemporary Readings*, Blackwell Publishers Ltd, Oxford, 1997.

Berners-Lee, Tim, *Weaving the Web: The Past, Present and Future of the World Wide Web by Its Inventor*, Orion Business, London, 1999.

Bodanis, David, *The Secret Family*, Simon & Schuster, New York, 1997.

Borges, Jorge Luis, *Labyrinths: Selected Stories and Other Writings*, New Directions, New York, 1964.

Braitenberg, Valentino, *Vehicles: Experiments in Synthetic Psychology*, MIT Press, Cambridge, Mass., 1994.

Brillat-Savarin, Jean-Anthelme, *The Physiology of Taste*, Penguin Classics, London, 1994.

Bruce, Vicki, and Young, Andy, *In The Eye of the Beholder: The Science of Face Perception*, Oxford University Press, Oxford, 1998.

Butler, Samuel, *Erewhon*, Penguin, London, 1985.

Bynum, W. F., and Porter, Roy (eds.), *Medicine and the Five Senses*, Cambridge University Press, Cambridge, 1993.

Calvin, William H., *The Cerebral Code: Thinking a Thought in the Mosaics of the Mind*, MIT Press, Cambridge, Mass., 1996.

Carter, Rita, *Mapping the Mind*, Weidenfeld & Nicolson, London, 1998.

Channell, David F., *The Vital Machine: A Study of Technology and Organic Life*, Oxford University Press, New York, 1991.

Cheney, Margaret, *Tesla: Man Out of Time*, Dorset Press, New York, 1989.

Classen, Constance, *The Color of Angels: Cosmology, Gender and the Aesthetic Imagination*, Routledge, London and New York, 1998.

Classen, Constance, Howes, David, and Synnott, Anthony (eds.), *Aroma: The Cultural History of Smell*, Routledge, London, 1994.

Crystal, David, *Language Death*, Cambridge University Press, Cambridge, 2000.

Cytowic, Richard E., *Synesthesia: A Union of the Senses*. Springer-Verlag, New York, 1989.

——, *The Man Who Tasted Shapes*, Abacus, London, 1993.

Dahl, Roald, *Charlie and the Chocolate Factory*, Puffin Books, London, 1995.

Damasio, Antonio R., *Descartes' Error: Emotion, Reason and the Human Brain*. Papermac, London, 1996.

——, *The Feeling of What Happens: Body, Emotion and the Making of Consciousness*, William Heinemann, London, 2000.

Darwin, Charles, *The Expression of the Emotions in Man and Animals*, with an introduction, afterword and commentaries by Paul Ekman, HarperCollins Publishers, London, 1998.

Davis, Erik, *Techgnosis: Myth, Magic and Mysticism in the Age of Information*, Serpent's Tail, London, 1998.

Davis-Floyd, Robbie, and Arvidson, P. Sven (eds.), *Intuition: The Inside Story*, Routledge, London, 1997.

Dennett, Daniel C., *Kinds of Minds: Towards An Understanding of Consciousness*, Phoenix, London, 1997.

Dertouzos, Michael, *What Will Be: How the New World of Information Will Change Our Lives*, Piatkus Publishers Ltd, London, 1997.

Dyson, George, *Darwin Among The Machines*, Penguin, London, 1997.

Edelman, Gerald M., *Bright Air, Brilliant Fire*, Penguin Books, London, 1992.

—— and Tononi, Giulio, *Consciousness: How Matter Becomes Imagination*, Allen Lane, the Penguin Press, London, 2000.

Ekman, Paul, *Telling Lies: Clues to Deceit in the Marketplace, Politics and Marriage*, W. W. Norton & Company, Inc., New York and London, 1992.

Gelernter, David, *The Muse in the Machine: Computerizing the Poetry of Human Thought*, the Free Press, New York, 1994.

Gershenfeld, Neil, *When Things Start To Think*, Hodder & Stoughton, London, 1999.

Grand, Steve, *Creation: Life and How To Make It*, Weidenfeld & Nicolson, London, 2000.

Gravelle, Karen, and Rivlin, Robert, *Deciphering the Senses: The Expanding World of Human Perception*, Simon and Schuster, New York, 1984.

Gray, Chris Hables (ed.), *The Cyborg Handbook*, Routledge, London, 1995.

Greenfield, Susan, *The Human Brain: A Guided Tour*, Weidenfeld & Nicolson, London, 1997.

Hillis, W. Daniel, *The Pattern on the Stone: The Simple Ideas that Make Computers Work*, Basic Books, New York, 1999.

Hodges, Andrew, *Alan Turing: The Enigma*, Vintage, London, 1992.

Houis, Jacques, Mieli, Paola, and Stafford, Mark (eds.), *Being Human: The Technological Extensions of the Body*, Agincourt/Marsilio, New York, 1999.

Howes, David (ed.), *The Varieties of Sensory Experience: A Sourcebook in the Anthropology of the Senses*, University of Toronto Press, Toronto, 1991.

Jeffrey, Mark, *The Human Computer*, Warner Books, London, 2000.

Jenkins, J. M., Oatley, K., and Stein, N. L., *Human Emotions: A Reader*, Blackwell Publishers, London, 1998.

Johnston, Victor S., *Why We Feel: The Science of Human Emotions*, Perseus Books, Reading, Mass., 1999.

Kottler, Jeffrey A., *The Language of Tears*, Jossey-Bass Publishers, San Francisco, 1996.

Kurzweil, Ray, *The Age of Spiritual Machines*, Orion Business Books, London, 1999.

Leary, Timothy, *Design for Dying*, with R. U. Sirius. HarperCollins Publishers, London, 1997.

LeDoux, Joseph, *The Emotional Brain*, Weidenfeld & Nicolson, London, 1998.

Le Guerer, Annick, *Scent: The Mysterious and Essential Powers of Smell*, Chatto & Windus, London, 1993.

Luria, A. R., *The Mind of Mnemonist*, Basic Books, New York, 1968.

Mazlish, Bruce, *The Fourth Discontinuity: The Co-evolution of Humans and Machines*, Yale University Press, New Haven and London, 1993.

McCrone, John, *Going Inside: A Tour Round a Single Moment of Consciousness*, Faber and Faber, London, 1999.

McLuhan, Marshall, *Understanding Media: The Extensions of Man*, MIT Press, Cambridge, Mass., 1997.

Miller, Jonathan, *The Body In Question*, Pimlico, London, 2000.

Mithen, Steven, *The Prehistory of the Mind: A Search for the Origins of Art, Religion and Science*, Phoenix, Weidenfeld & Nicolson, London, 1996.

Montaigne, Michel de, *The Complete Essays*, translated by M. A. Screech, Penguin Books, London, 1991.

Moravec, Hans, *Mind Children: The Future of Robot and Human Intelligence*, Harvard University Press, Cambridge, Mass., 1988.

——, *Robot: Mere Machine to Transcendent Mind*, Oxford University Press, Oxford, 1999.

Nabokov, Vladimir, *Speak, Memory*, Perigree Books, New York, 1979.

Napier, John, *Hands*, revised by Russel H. Tuttle, Princeton University Press, Princeton, 1993.

Norretranders, Tor, *The User Illusion: Cutting Consciousness Down to Size*, Penguin, London, 1998.

Nyberg, David, *The Varnished Truth: Truth Telling and Deceiving in Ordinary Life*, University of Chicago Press, Chicago, 1994.

Marinetti, Filippo Tommaso, *The Futurist Cookbook*, Trefoil Publications, London, 1989.

Panksepp, Jaak, *Affective Neuroscience: The Foundations of Human and Animal Emotions*, Oxford University Press, Oxford, 1998.

Park, David, *The Fire Within the Eye: A Historical Essay on the Nature and Meaning of Light*, Princeton University Press, Princeton, 1997.

Penrose, Roger, *The Emperor's New Mind*, Oxford University Press, Oxford, 1999.

Pinker, Steven, *The Language Instinct: The New Science of Language and Mind*, Allen Lane, the Penguin Press, London, 1994.

——, *How The Mind Works*, Allen Lane, the Penguin Press, London, 1998.

Picard, Rosalind, *Affective Computing*, the MIT Press, Cambridge, Mass., 1997.

Proust, Marcel, *Remembrance of Things Past*, vol. I, translated by C. K. Scott Moncrieff and Terence Kilmartin, Random House, New York, 1981.

Ramachandran, V. S., and Blakeslee, Sandra, *Phantoms in the Brain*, Fourth Estate, London, 1999.

Rimbaud, Arthur, *Complete Works*, translated by Paul Schmidt, Harper & Row Publishers, New York, 1976.

Rose, Steven, *The Making of Memory: From Molecules to Mind*, Bantam Books, London, 1995.

Sachs, Oliver, *The Man Who Mistook His Wife for a Hat*, Duckworth, London, 1987.

Searle, John R., *The Mystery of Consciousness*, Granta Books, London, 1997.

Schrödinger, Erwin, *What Is Life? with Mind and Matter and Autobiographical Sketches*, Canto, Cambridge, 2000.

Shear, Jonathan (ed.), *Explaining Consciousness: The Hard Problem*, the MIT Press, Cambridge, Mass., 1997.

Squire, Larry R., and Kandel, Eric R., *Memory: From Mind to Molecules*, Scientific American Library, New York, 1999.

Süskind, Patrick, *Perfume*, Penguin Books, London, 1987.

Tesla, Nikola, *My Inventions*, Barnes & Noble Books, New York, 1995.

Turkle, Sherry, *Life on the Screen: Identity in the Age of the Internet*, Phoenix, London, 1995.

Underhill, Paco, *Why We Buy: The Science of Shopping*, Orion Business Books, London, 1999.

Von Neumann, John, *The Computer and The Brain*, Yale University Press, New Haven and London, 2000.

Vroon, Piet, *Smell: The Secret Seducer*, Farrar, Straus and Giroux, New York, 1997.

Warde, Mark, *Virtual Organisms: The Startling World of Artificial Life*, Pan Books, London, 1999.

Warwick, Kevin, *March of the Machines: Why the New Race of Robots Will Rule the World*, Century, London, 1997.

Watson, Lyall, *Jacobson's Organ and the Remarkable Nature of Smell*, Allen Lane, the Penguin Press, London, 1999.

Whitman, Walt, *Complete Poetry and Selected Prose*, Houghton Mifflin Company, Boston, 1959.

Wiener, Norbert, *The Human Use of Human Beings: Cybernetics and Society*, Da Capo Press, Inc., New York, 1954.

——, *God & Golem, Inc.: A Comment on Certain Points where Cybernetics Impinge on Religion*, the MIT Press, Cambridge, Mass., 1990.

Wills, Garry, *Saint Augustine*, Weidenfeld & Nicolson, London, 1999.

Wurm, Stephen, *Atlas of the World's Languages in Danger of Disappearing*, UNESCO Publishing/Pacific Linguistics, Paris/Canberra, 1996.

Yates, Frances A., *The Art of Memory*, Pimlico, London, 1999.

INDEX

Page references in **bold** indicate illustrations.